APPLICATIONS AND EFFICIENCY OF HEAT PUMP SYSTEMS

sti

ACKNOWLEDGEMENTS

The valuable assistance of the Technical Advisory Committee and panel of referees is gratefully acknowledged.

TECHNICAL ADVISORY COMMITTEE

Professor I E Smith (Chairman)	Cranfield Institute of Technology
Mr R Benstead	The Electrical Council Research Centre
Dr T Heppenstall	NEI International Research & Development Co Ltd
Professor F A Holland	University of Salford
Dr R King	BHR Group Ltd
Dr R L Powell	ICI Mond Division
Mr P Welsby	British Gas plc
Miss J Thorn (Conference Organiser)	BHR Group Ltd

OVERSEAS CORRESPONDING MEMBERS

Professor G Alefeld	Technische Universität München	Germany
Professor T Berntsson	Chalmers University of Technology	Sweden
Dr P Coda	ENEA	Italy
Mr D C Erickson	Energy Concepts Co	USA
Mr J A Knobbout	Consultant	The Netherlands
Professor P Worsoe-Schmidt	Technical University of Denmark	Denmark

APPLICATIONS AND EFFICIENCY OF HEAT PUMP SYSTEMS

Proceedings of the
4th International Conference
(Munich, Germany 1-3 October 1990)

I E Smith (Editor)
Cranfield Institute of Technology, UK

Springer-Verlag Berlin Heidelberg GmbH
Organised and sponsored by BHR Group Limited

First published in 1991 by:
STI
4 Kings Meadow
Ferry Hinksey Road
Oxford
OX2 0DU
Tel: (0865) 798898 Fax: (0865) 798788

and

Originally published by STI, Springer-Verlag, New York in 1991

British Library Cataloguing in Publication Data
Applications and efficiency of heat pump systems: proceedings of the 4th. international Conference
1. Heat pumps
621.4025

ISBN 978-3-662-30181-4 ISBN 978-3-662-30179-1 (eBook)
DOI 10.1007/978-3-662-30179-1

When citing papers from this volume, the following reference should be used: Title, Author(s), Paper Name, Pages, Proceedings of the International Conference on Heat Pumps, Munich, Germany. Organised by BHR Group (1-3 October 1990).

CONTENTS

FOREWORD

The papers in this collection have originated from Britain, Eastern and Western Europe and India, with the delegates coming from fifteen countries, including a strong contingent from Japan. This indicates a widespread interest in the application of heat pumps.

The heat pump suffers from an environmental dichotomy. On the one hand it saves fossil fuel energy, thereby reducing CO_2 emissions. On the other hand, in the vapour compression form, it generally employs CFCs which are destructive to our protective ozone layer as well as contributing to the greenhouse effect.

Taking the first, heat pumps, perhaps have the widest application numerically in the heating (and cooling) of buildings and an excellent paper describes case studies concerning three large Norwegian hotels. In these, heat was pumped from the adjacent river or sea when heating was required, and during the summer, when cooling became necessary, to generate domestic hot water. The heat pumps were installed by SINTEF Refrigeration Engineering, Norway, and have demonstrated payback periods of about two years. The fractional total energy saving of the three hotels was 30% as a result, an impressive figure, indeed. A similar paper by a Belgian architectural consortium shows how this technique can be successfully applied to the cooling of a large television complex where considerable quantities of heat were being generated by the luminaries. In this exercise the cooling load was successfully pumped to provide hot water. A third paper examines the viability of solar collectors to provide a heat source suitable for pumping heat into an army barracks complex in West Germany.

In light of such evidence one wonders why the more general application of heat pumps on the commercial scale has yet to materialise. One may hypothesise that the building services industry has yet to accept them as being the norm and so heat pumps remain in gestation.

Turning to the other side of the coin and addressing the pollution effects of the CFC gases presently used in heat pumping and refrigeration an interesting paper by a group from Graz University, Austria, suggests the possibility of using Ethyl Chloride. This fluid has an extremely short atmospheric lifetime, is well known and understood, and its sole disadvantage is in its flammability.

One of the main factors behind the use of CFCs in refrigeration has been their very desirable inflammability and low toxicity, but questions are raised as to whether non-flammability is a necessary criterion. Given a relaxation on this would open up a wide range of available refrigerants, the penalty being the upgrading of the total installation to flameproof standards.

However, in process industries already handling inflammable materials, for example oil products and solvents, such a requirement would impose no additional burden since such plants would already be flame proofed.

If non-flammability remains a necessary criterion refrigerants R134 and 134a are at present the most promising replacements and ICI are making progress in their commercial realisation. At present they are undergoing examination for toxicity, necessarily a lengthy process, and satisfactory lubricants for these fluids have yet to be proven.

A different approach providing a partial solution to the CFC problem comes from the Chalmers University, Sweden suggesting that by employing a mixture of a CFC with an environmentally-friendly (although flammable) refrigerant which was non-flammable might be an acceptable compromise. An ingenious solution, certainly, but one which this writer feels must be relegated to the 'brave attempt' category. The refrigeration industry has been very severely shaken by the Montreal Convention and its implications for them, and now needs a completely fire-proof solution. It is hardly likely to accept any half-measure vulnerable to further attack in the future.

Professor I E Smith
October 1990

PART I

System design

Chapter 1

DESIGN OF A VAPOUR COMPRESSION HEAT PUMP IN A HEAT EXCHANGER NETWORK

V R Pendyala and V S Patwardhan (National Chemical Laboratory, India)

ABSTRACT

It is known that the proper placement of a vapour compression heat pump in a heat exchanger network (HEN) is across the pinch. In this paper a detailed procedure is developed for designing such a heat pump. Design based on the composite curves is first considered. The modification to be used if only selected streams are to be routed to the heat pump is also discussed. The possibility of another pinch developing in the remaining HEN problem is taken into account. Illustrative examples are presented at the end which bring out the details clearly. The present procedure is a useful tool for a designer for designing a HEN with a heat pump from the point of view of overall economics.

NOMENCLATURE

$(COP)_A$	actual coefficient of performance ($=Q_{CO}/W$)	
$(COP)_R$	Rankine coefficient of performance	
FC_p	heat capacity flow rate	(kW/^oC)
Q	heat load	(kW)
$Q_1 - Q_3$	refer to Fig.6	
T	temperature	(^oC)
$T_0 - T_3$	refer to Fig.6	
ΔT_{min}	minimum temperature driving force	(^oC)
W	high grade energy	(kW)

Subscripts

A - I	refer to Fig.2
c	cold composite or cold streams
CO	condenser
EV	evaporator

Heat Pumps – Proceedings of the 4th International Conference, pp.3-18

h	hot composite or hot streams
s	source
T	target

Greek letters

α	heat balance error(eq(2))
β	heat pump effectiveness factor

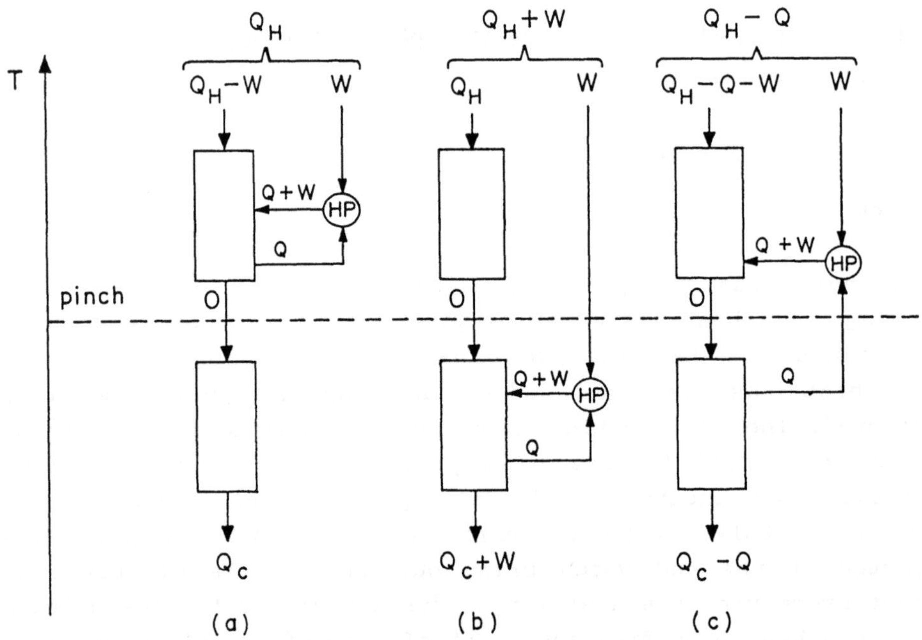

Fig. 1. Placement of a heat pump in a HEN

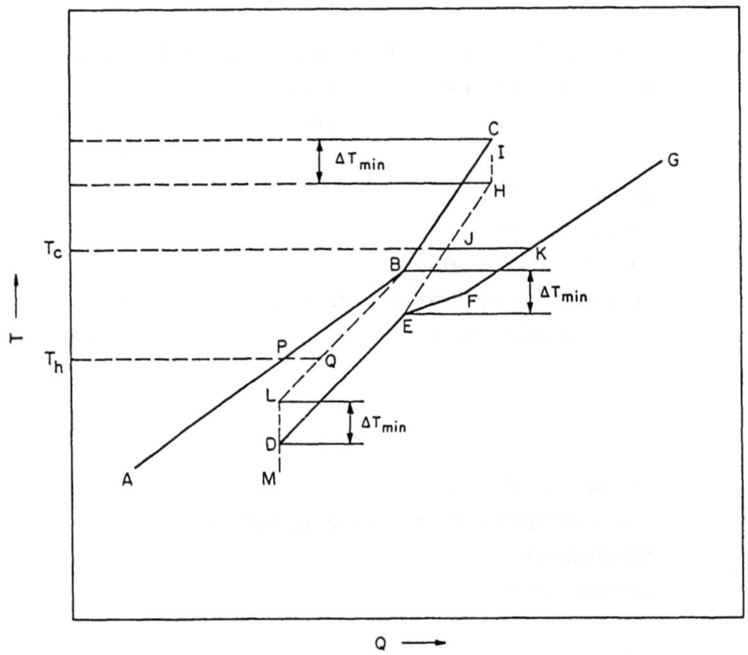

Fig. 2. Typical composite diagram

INTRODUCTION

Heat pumps are devices which upgrade low level heat energy to a higher, more useful temperature level by consuming a relatively small amount of high grade energy. Process integration is a technique whereby the maximum amount of heat energy can be interchanged within a process, thereby reducing utility consumption. Both these concepts essentially result in a reduction of the primary energy requirement of a process. Although both these concepts have been extensively investigated separately, their interaction with each other has been investigated only rather recently. Townsend and Linnhoff (1983) were the first to show clearly that the two concepts are not contradictory or mutually exclusive, but rather, they can both be used simultaneously, with advantage, in certain types of problems. They also explicitly showed that the appropriate placement of a vapour compression heat pump is across the pinch for realising its full benefit. Its placement above the pinch is equivalent to replacing part of the hot utility by the more expensive electricity and its placement below the pinch is equivalent to converting electricity into cooling water. This is shown schematically in Fig.1. A procedure for designing a heat pump based on the composite curves was developed later (Ranade et al., 1986a, 1986b) which describes how to arrive at the proper size and temperature levels for a heat pump for the maximum recycle of heat energy. A question that has not been analysed in detail so far is the possibility of using only selected streams for the heat pump, and its relationship with the size and temperature levels of the heat pump. The question of multiple utilities has also been discussed more recently in the context of heat pumps (Ranade et al., 1987).

In this paper, a detailed procedure is developed for designing a heat pump across the pinch, ensuring that the capacity is the maximum that is consistent with non-violation of the pinch condition anywhere in the process. This procedure is different from the one referred to earlier, and is based on the conceptual removal of streams from the composite diagrams. This makes it possible to select certain streams specifically for being routed to the heat pump if so desired. It also takes into account the possibility of different working fluids being used, the exact dependence of $(COP)_A$ on working temperature levels and expected heat losses from the heat pump circuit. Some results of calculation are presented for the purpose of illustration. These results were obtained using the software that has been developed, which takes stream data, generates composite diagrams, locates the pinch, and suggests an appropriate working fluid that would be suitable. At this stage, the user can, in an interactive mode, select another working fluid if so desired. The software then uses thermodynamic properties of the selected working fluid, designs the largest possible heat pump, and suggests the streams to be routed to the heat pump for implementing the same. At this stage, the user can indicate specific streams to be used and the software comes up with the largest possible heat pump using only these streams. It may be noted that the temperature levels get recalculated automatically. The software has a built-in facility which allows the user to try different options for various parameters such as the working fluid, streams removed for use in the heat pump circuit etc. till he is satisfied with the results.

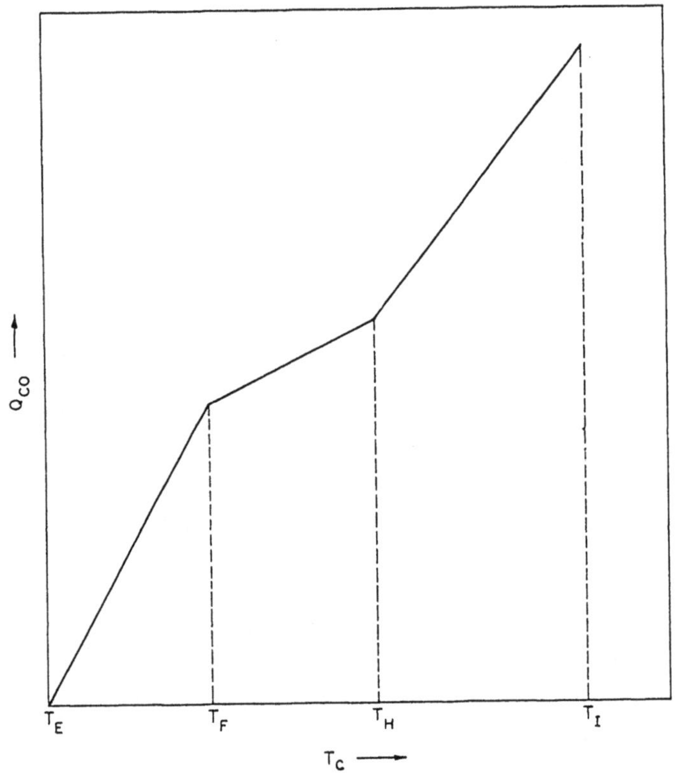

Fig. 3. $Q_{CO} - T_C$ diagram from Fig. 2

THEORY

The construction of a composite diagram from given stream data (source and target temperatures, and heat capacity flow rates) is well known and need not be repeated here (for example, see Linnhoff et al., 1982). Figure 2 shows a typical pair of hot and cold composite curves which forms the basis for the following discussion. More often than not, only one of the composite curves changes slope at the pinch. Figure 2 shows both of them changing slopes at the pinch, which is merely a general representation. The following discussion is equally valid in both these cases. Line ABC represents the hot composite while line DEFG the cold composite. Points B and E represent the temperature pinch. It has been stated earlier that a heat pump in such a situation must receive heat from below the pinch and deliver it above the pinch. Consider first, the delivery of heat above the pinch. Line EH in

Fig.2 has been obtained by shifting line segment BC (the hot composite above the pinch) downwards by ΔT_{min}. Line HI is a vertical line. Consider any point K on line EFG (which is the cold composite above the pinch). If any cold stream above the pinch is removed and rerouted to the heat pump condenser, the composite diagram of the remaining cold streams (above the pinch) will be shifted to the left. Point K can thus be shifted to the left, all the way up to J. If it is shifted further to the left, it will violate the pinch condition. Thus, enthalpy JK represents the maximum load, Q_{CO} that can be diverted to the heat pump condenser without violating the pinch condition in the remaining problem, corresponding to the temperature T_c. By plotting Q_{CO} vs. T_c, we get Fig.3 which shows the excess enthalpy of the cold composite above the pinch. T_I of course is the maximum value of T_c up to which heat can be added to the cold composite.

Let us now consider removal of heat from the hot composite below the pinch. Line BL in Fig.2 has been obtained by shifting ED upwards by ΔT_{min}, and LM is a vertical line. Enthalpy PQ represents the maximum load, Q_{EV}, that can be diverted to the heat pump evaporator without violating the pinch condition, corresponding to temperature T_h. Figure 4 shows the Q_{EV} vs. T_h curve obtained by varying T_h. It may be noted that Fig. 3 and Fig. 4 are, in fact, sections of the grand composite curve defined by Linnhoff et al. (1982) with the axes interchanged.

Design of a Heat Pump

Figures 3 and 4 form the starting point for sizing a heat pump that is to operate across the pinch. If these curves are not monotonic, they need to be smoothened out before they can be used for this purpose, as discussed later. These figures give the Q_{CO} - T_c and Q_{EV} - T_h relationships. In addition, the following relationships are valid :

(i) The actual coefficient of performance, $(COP)_A$ which is equal to (Q_{CO} / W) is given by

$$(COP)_A = \beta (COP)_R \qquad \text{---- (1)}$$

where β is the heat pump effectiveness factor.

(ii) The heat balance error (which accounts for heat losses from the system) is defined as

$$\alpha = (-Q_{CO} + Q_{EV} + W) / Q_{CO} \qquad \text{---- (2)}$$

Elimination of W gives

$$Q_{CO} = Q_{EV} / (1 + \alpha - 1/(COP)_A) \qquad \text{---- (3)}$$

The Rankine coefficient of performance can be written as

$$(COP)_R = f (T_{CO}, T_{EV}) \qquad ---- (4)$$

where

$$T_{CO} = T_c + \Delta T_{min} \qquad ---- (5)$$

$$T_{EV} = T_h - \Delta T_{min} \qquad ---- (6)$$

and the function f depends not only on T_{CO} and T_{EV}, but on the working fluid being used as well.

The heat pump effectiveness factor and the heat balance error can be estimated based on experience with heat pump systems (for example, see Gutierrez et al., 1979). Also from economic considerations, it is usually possible to fix some reasonable value of $(COP)_A$ for the heat pump from known utility values, capital costs and other economic factors. Using these values, the seven parameters, namely, $(COP)_R$, Q_{CO}, Q_{EV}, T_{CO}, T_{EV}, T_c and T_h can be calculated using equations (1), (3) - (6) and Fig.3 and 4. The specific procedure used in this work was to assume a trial value of T_h, get T_{EV} from eq(6), Q_{EV} from Fig.4, Q_{CO} from eq(3), T_c from Fig.3, T_{CO} from eq(5), $(COP)_R$ from eq(4) and $(COP)_A$ from eq(1). This $(COP)_A$ was compared to the $(COP)_A$ required based on economic considerations and the next trial value was selected for T_h. The half interval method (Carnahan et al., 1969) can be conveniently used for converging to the correct value of T_h.

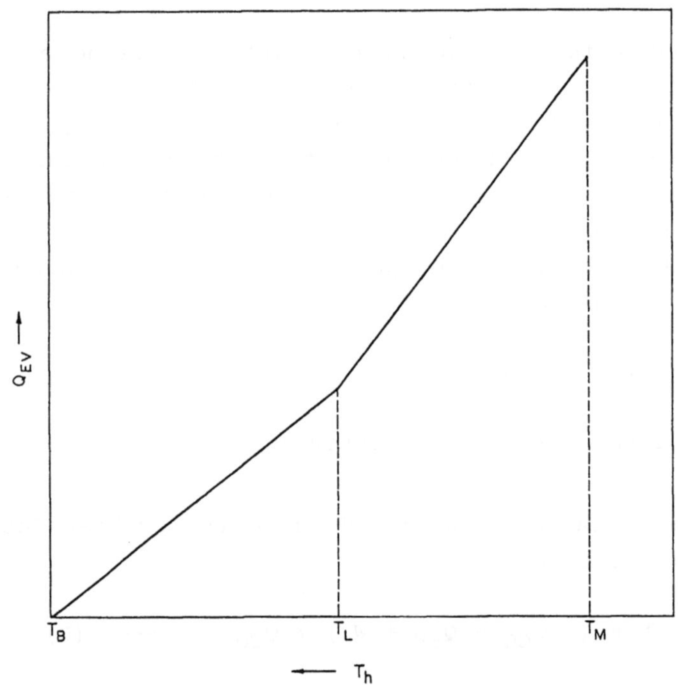

Fig. 4. $Q_{EV} - T_h$ diagram from Fig. 2

The Possible Development of Second Pinch

The discussion so far was based on the composite curves of Fig.2, which gave monotonically increasing curves in Fig.3 and 4. Here we discuss the case where one or both of the Q_{CO} - T_c and Q_{EV} – T_h curves go through a maximum. Consider a problem whose composite curves are given by Fig.5. Following the procedure outlined in the earlier section, we get the Q_{CO} - T_c curve shown in Fig.6 (RSTUV). If cold streams having enthalpy Q_1 in the interval of T_1 - T_0 are removed (i.e. routed to the heat pump condenser) from the problem, the composite diagrams for the remaining problem shows violation of the ΔT_{min} condition at T_2. Thus, the maximum enthalpy that can be routed to the heat pump condenser is Q_2 up to any temperature $T_c < T_2$. This can be incorporated in the Q_{CO} - T_c diagram simply by removing maxima in Fig.6 (RSUV). Thus, all the maxima in the Q_{CO} - T_c diagram need to be removed before following the design procedure described in the earlier section. It is also obvious that maxima from the Q_{EV} - T_h curve need to be removed as well. The modified curves can then be used for designing a heat pump. Smoothening of several maxima is illustrated in Fig.7 for an imaginary Q_{EV} - T_h curve (where the curve ABCDEFG is smoothened to get ABDEG). It may be noted that Q_{CO} and Q_{EV} cannot be greater than the minimum hot and cold utility requirements respectively.

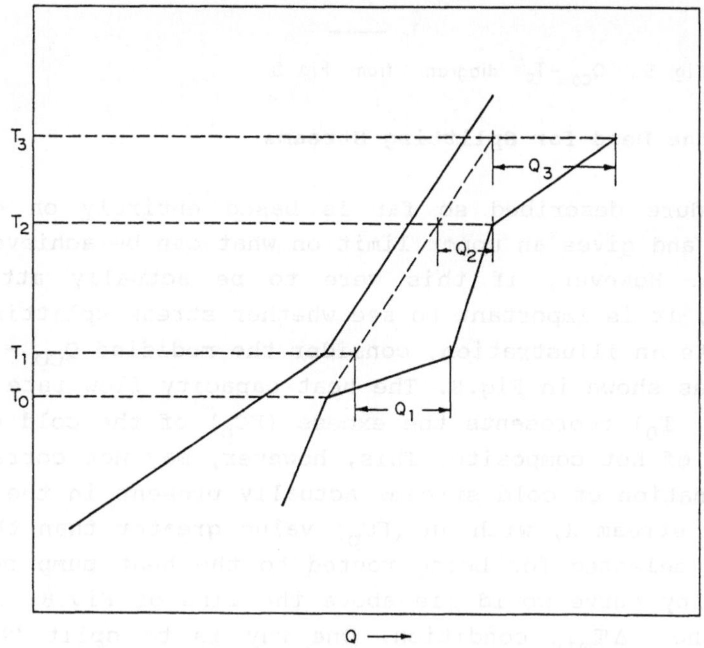

Fig. 5. Composite diagrams with possible second pinch

At first it may appear that if the use of a heat pump leads to the developoment of another pinch, one could use a second heat pump there. However, this is not possible as it leads to a violation of the ΔT_{min} constraint at the first pinch. If the problem has two pinches, energy recycling is possible only if the heat pump takes heat from a point below both the pinches and delivers it above both of them. From a practical point of view, such a heat pump would be uneconomical due to the large temperature lift involved in most cases.

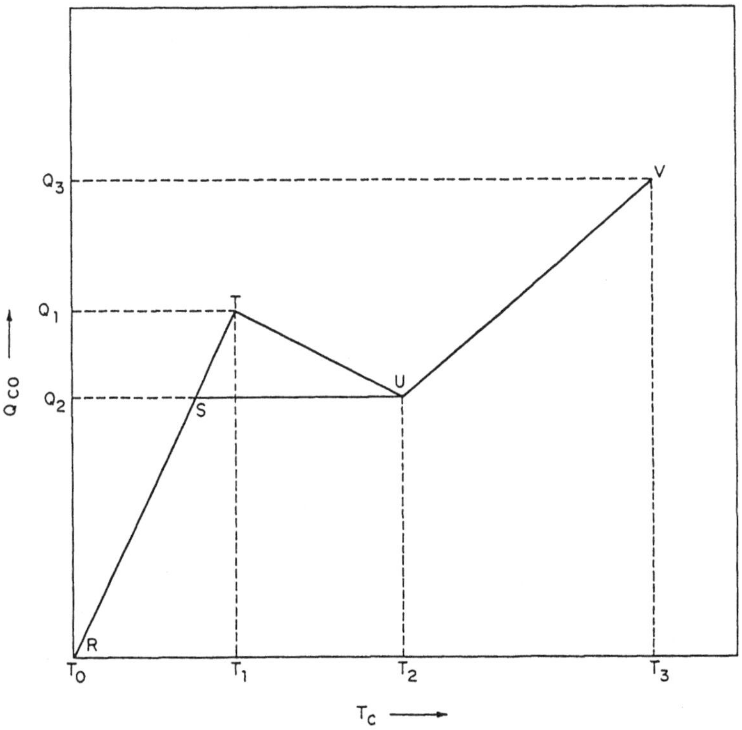

Fig. 6 . $Q_{co} - T_c$ diagram from Fig. 5

Avoiding the Need for Splitting Streams

The procedure described so far is based entirely on composite diagrams, and gives an upper limit on what can be achieved with a heat pump. However, if this were to be actually attained in operation, it is important to see whether stream splitting can be avoided. As an illustration, consider the modified $Q_{CO} - T_c$ curve of Fig.6 as shown in Fig.8. The heat capacity flow rate given by $Q_2/(T_1 - T_0)$ represents the excess (FC_p) of the cold composite over that of hot composite. This, however, may not correspond to any combination of cold streams actually present in the problem. If a cold stream X, with an (FC_p) value greater than the excess (FC_p), is selected for being routed to the heat pump condenser, its enthalpy curve would lie above the line of Fig.8, and would violate the ΔT_{min} condition. One way is to split the stream appropriately and route only a portion of it to the heat pump. There is a better way however. Figure 8 shows the line T_X - S which is parallel to the line for cold stream X. To avoid stream splitting we could leave stream X in the HEN till it attains the temperature of T_X and then route it to the heat pump condenser where it would get heated to T_4. This way, we can avoid stream splitting. If the selected stream has an (FC_p) value less than the excess (FC_p) (see the line of cold stream Y in Fig.8), then splitting is not required. It can be heated to the temperature T_Y in the heat pump condenser. The designer of course has the choice of selecting one or several streams to be routed to the heat pump such that their total enthalpy - temperature diagram lies below the modified Q_{CO} - T_c curve of Fig.8. Once the streams have been selected, a Q_{CO} - T_c diagram can be constructed for them, which becomes the basis for designing a heat pump. Similar comments hold below the pinch as well. This procedure is clearly illustrated in the following examples.

Handling limiting values of stream temperatures

The Q_{CO} - T_c and Q_{EV} - T_h curves for selected streams have upper or lower limit on temperatures as well as heat loads. This can be due to several reasons such as the target (or source) temperature and FC_p values of the selected streams, limiting temperatures on the composite diagrams, etc. In such cases, the limiting temperature can be selected, and the half interval method mentioned above can be used for converging to the correct value of $(COP)_A$ which in such cases is higher than the economically selected minimum value. The details can be seen in the illustrative examples.

ILLUSTRATIVE EXAMPLES

The procedure presented above is illustrated in this section by applying it first to the example problem involving four streams with the stream data given in Table I. The composite diagrams for the hot and cold streams are shown in Fig.9. The minimum utility requirements without a heat pump are 107 kW (hot) and 64 kW (cold). The pinch occurs at a temperature of 150°C (hot) and 140°C (cold). In order to operate across the pinch, a heat pump

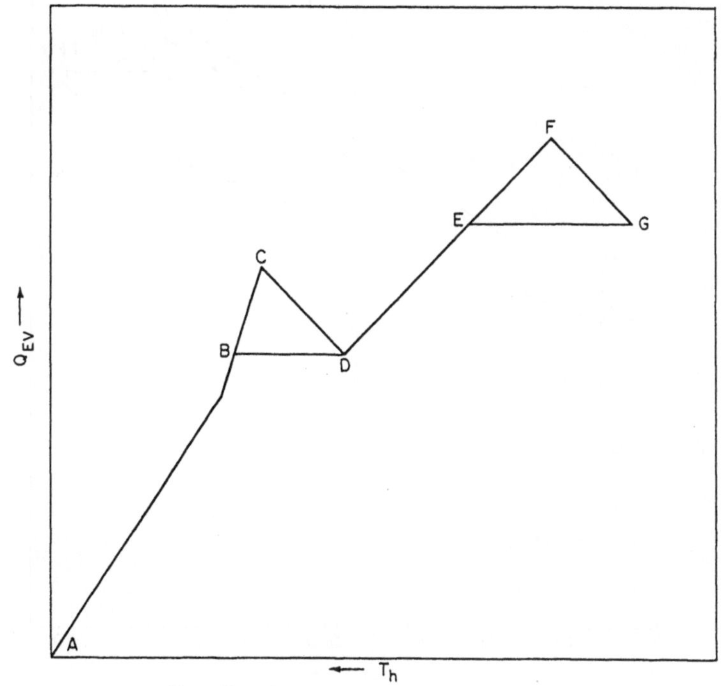

Fig. 7. Smoothing of multiple maxima

would have to operate with a condensing temperature much greater than 140°C. Till a few years ago, such heat pumps would have been considered impractical. The situation has, however, changed with the availability of steam compressors. The Q_{CO} - T_c plot shown in Fig.10 (dotted line) shows a maximum and needs to be smoothed out into a monotonic curve (the continuous line). The Q_{EV} - T_h plot is shown in Fig.11.

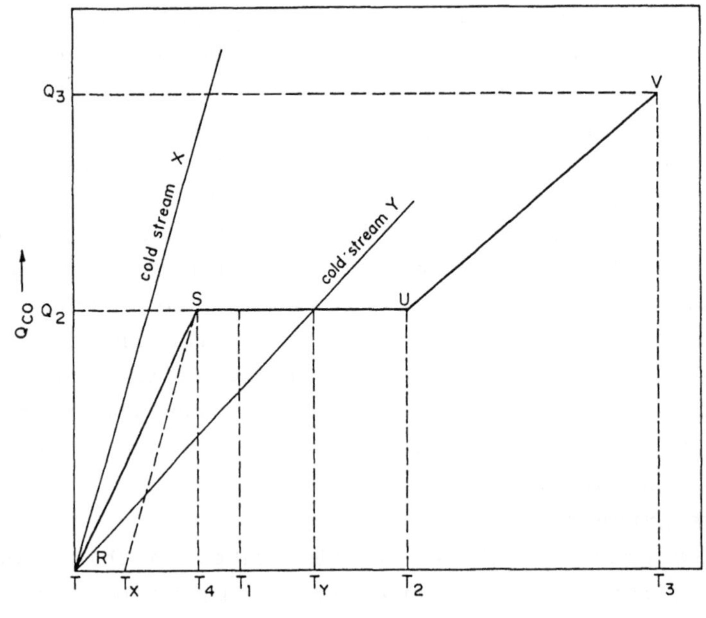

Fig. 8. Use of $Q_{CO} - T_c$ diagram for heat pump design

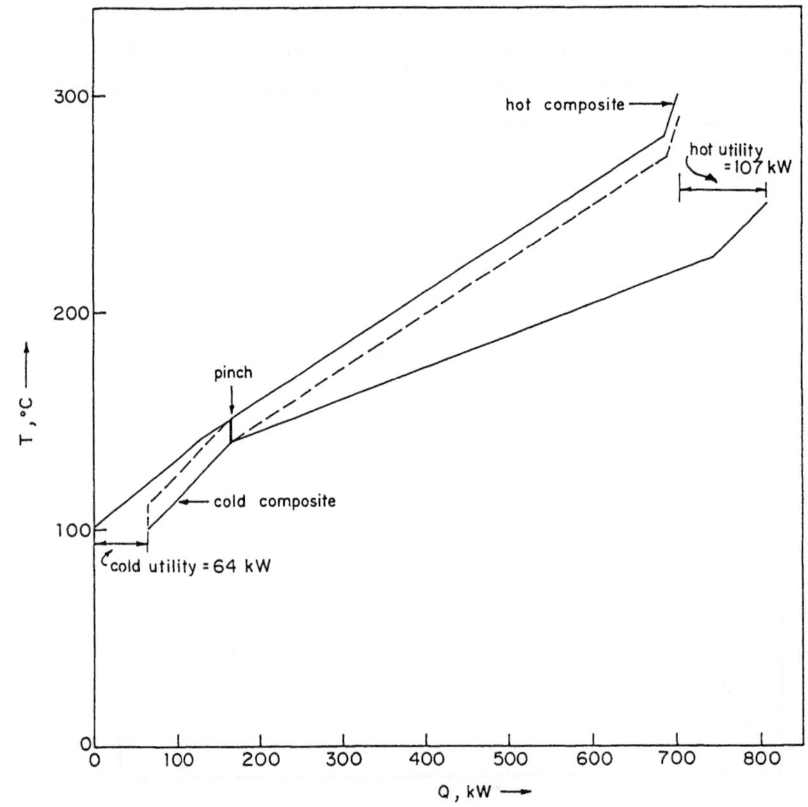

Fig. 9. Composite diagram for example I

For designing a heat pump, the parameter values used are : $\beta = 0.65$, $\alpha = 0.0$ and $(COP)_A = 4.0$. The working fluid selected was R718 (namely, steam) which is thermodynamically suitable for these temperature levels. These parameter values have been

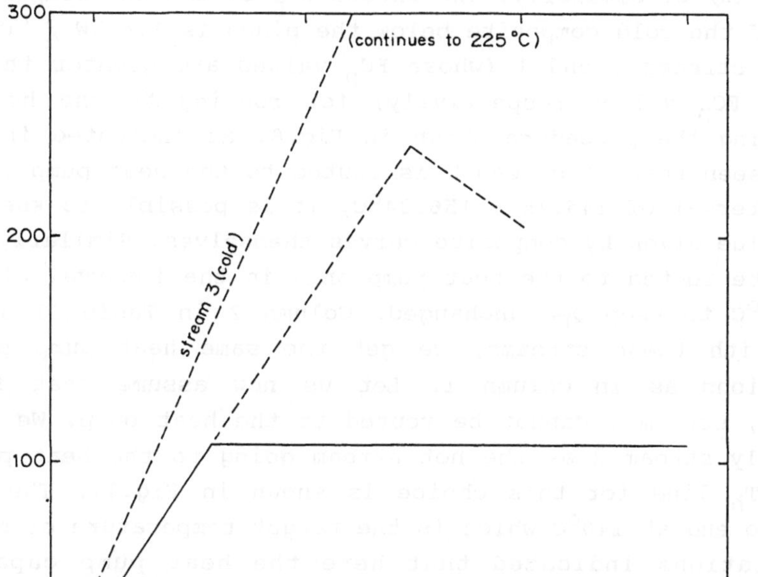

FIG.10 : $Q_{CO}-T_C$ DIAGRAM FOR EXAMPLE I

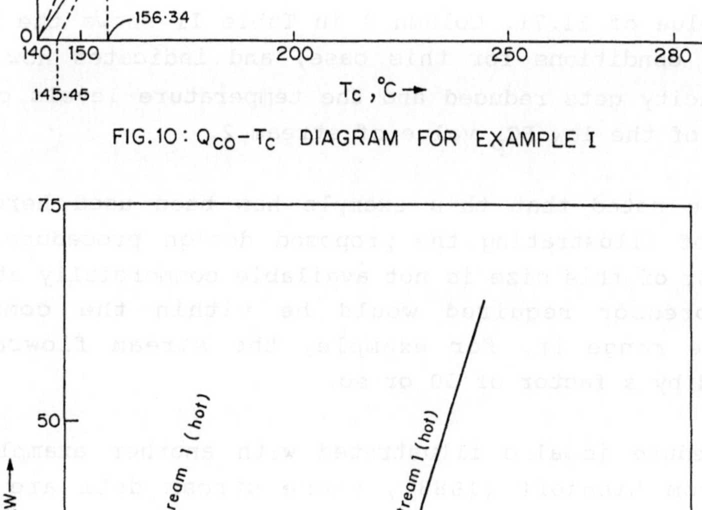

Fig. 11. $Q_{EV}-T_h$ diagram for example I

selected as reasonable and mainly for the purpose of illustration. The resulting heat pump parameters are shown in Table II (first column). It may be noted that the (FC_p) value of the cold composite in excess of the hot composite above the pinch is 2.8 kW/^{o}C. Similarly, the excess FC_p of the hot composite over that of the cold composite below the pinch is 1.4 kW / ^{o}C. Let us select streams 3 and 1 (whose FC_p values are greater than these excess FC_p values respectively) for routing to the heat pump. Following the procedure shown in Fig.8, as indicated in Fig.10, it is seen that if stream 3 is routed to the heat pump only over the interval of 145.45 - 156.34^{o}C, it is possible to keep Q_{CO} at the value given by composite curves themselves. Similarly, stream 1 can be routed to the heat pump only in the interval of 120.0 - 109.28^{o}C to keep Q_{EV} unchanged. Column 2 in Table II indicates that with these streams, we get the same heat pump operating conditions as in column 1. Let us now assume that for some reason, stream 1 cannot be routed to the heat pump. We can then use only stream 2 as the hot stream going to the heat pump. The Q_{EV} - T_h line for this choice is shown in Fig.11. The line is seen to end at 140°C which is the target temperature of stream 2. Calculations indicated that here the heat pump capacity is limited by the enthalpy available in this stream, which is equal to 8.0 kW. The procedure described earlier converged on the $(COP)_A$ value of 11.71. Column 3 in Table II shows the heat pump operating conditions for this case, and indicates how the heat pump capacity gets reduced and the temperature levels change, as a result of the low FC_p value of stream 2.

It may be noted that this example has been used here for the purpose of illustrating the proposed design procedure. A steam compressor of this size is not available commercially at present. The compressor required would be within the commercially available range if, for example, the stream flowrates were increased by a factor of 50 or so.

The procedure is also illustrated with another example problem taken from Linnhoff (1983), whose stream data are shown in Table III. The minimum utility requirements without a heat pump are 60 kW (hot) and 225 kW (cold). The pinch occurs at a temperature of 150°C (hot) and 140° (cold). The resulting heat pump parameters based on the composite diagram as well as selected streams are shown in Table IV. It may be noted that heat pump based on composite streams (as shown in column 1) is limited by the highest target temperature of cold streams (180°C) which limits T_{CO} to 190°C. The $(COP)_A$ is seen to be higher than the selected minimum value of 4.0. Column 2 shows that we get the same heat pump if stream 2 and 3 are selected. If stream 1, whose FC_p is less than the excess FC_p below the pinch, is selected for the heat pump then the capacity of the heat pump gets reduced as indicated by the result in column 3.

Table I : Stream data for example problem 1

No.	Type	FC_p kW/^oC	T_S ^oC	T_T ^oC
1	hot	3.2	280.0	100.0
2	hot	0.8	300.0	140.0
3	cold	4.2	140.0	225.0
4	cold	2.6	100.0	250.0

$\Delta T_{min} = 10.0^oC$

Table II : Heat pump parameters for different combination of streams

Parameter	Composite streams	Selected streams 1,3	2,3
T_{CO}, ^oC	166.34	166.34	153.12
T_{EV}, ^oC	99.28	99.28	130.00
Q_{CO}, kW	45.75	45.75	8.75
Q_{EV}, kW	34.31	34.31	8.00
W, kW	11.44	11.44	0.75
$(COP)_A$	4.00	4.00	11.71

Table III : Stream data for example problem 2

No.	Type	FC_p kW/^oC	T_S ^oC	T_T ^oC
1	hot	2.0	180.0	40.0
2	hot	4.0	150.0	40.0
3	cold	3.0	60.0	180.0
4	cold	2.6	30.0	105.0

$\Delta T_{min} = 10.0^oC$

Table IV : Heat pump parameters for different combinations of streams

Parameter	Composite streams	Selected streams 2,3	Selected streams 1,3
T_{CO}, ^{o}C	190.00	190.00	188.80
T_{EV}, ^{o}C	124.65	124.65	118.80
Q_{CO}, kW	60.00	60.00	56.50
Q_{EV}, kW	46.05	46.05	42.40
W, kW	13.95	13.95	14.10
$(COP)_A$	4.30	4.30	4.00

DISCUSSION

It has been mentioned in the previous section that for designing a heat pump across the pinch, some $(COP)_A$ value needs to be used. The lower this value, the higher are both the permissible temperature lift, and the heat pump capacity. However, it also implies a higher consumption of the high grade motive energy. Thus the value of $(COP)_A$ to be used is determined by economics. The exact formulation of an economic criterion for a heat pump operating in isolation is well established (Holland et al., 1982). In case of a heat pump in a heat exchanger network this becomes rather complex. It has been shown by Ranade et al. (1987) that the incorporation of a heat pump in a heat exchanger network usually leads to an increased area requirement for the remaining problem, due to the somewhat reduced temperature driving force. Any economic considerations that are applied to determine the minimum $(COP)_A$ applicable must take this into account. However, the exact implications of the increased area requirement can be calculated only when the HEN synthesis for the remaining problem is completed. (It may be borne in mind that the equation given by Ranade et al (1987) for the maximum economic temperature lift involves a term called "thermodynamic efficiency" which is a complex function of various parameters such as heat pump temperature levels etc. which makes their equation a highly implicit one.) The present procedure is thus really a tool in the hands of a designer who can start with a $(COP)_A$ value based on judgment, and correct it later if required.

The present procedure also describes how to avoid stream splitting during design of a heat pump, as it can be applied not only to composite diagrams but also to individually selected streams. In some problems it may turn out that even if stream splitting is avoided during design of a heat pump, the HEN for the remaining problem may involve stream splitting anyway. Why not incorporate it during heat pump design itself for such problems ? This again underlines the fact that the finally accepted design of a heat pump and HEN must come out of analysis of both these aspects in totality, depending upon both economic and operational considerations.

CONCLUSIONS

A heat pump can be designed to operate across the pinch in a HEN by conceptually removing streams from the HEN and routing them to the heat pump. The possible development of another pinch in the HEN may limit the heat pump capacity. If large heat capacity streams cannot be used in the heat pump, the amount of energy that can be recycled, can reduce considerably. The design procedure developed incorporates several options that can be used by a design engineer to come up with the most suitable design of a heat pump in a HEN.

REFERENCES

1. Carnahan, B., Luther, H. A. and Wilkes, J. O., 1969, "Applied numerical methods", John Wiley & Sons, New York, 178-180.

2. Colmenares, T. R. and Seider, W. D., 1989, "Synthesis of utility systems integrated with chemical processes", Ind. Eng. Chem. Res., Vol.28(1), 84-93.

3. Gutierrez, A. G., El-Meniawy, S. A. K., Watson, F. A. and Holland, F. A., 1979, "Heat pumps : Part V - operating characteristics of a water - water heat pump systems using R12", Ind. Chem. Eng., Vol.XXI(4), 16-26.

4. Holland, F. A., Watson, F. A. and Devotta, S., 1982, "Thermodynamic design data for heat pump systems", Pergamon Press, Oxford, 19-20.

5. Linnhoff, B. and Hindmarsh, E., 1983, "The pinch design method for heat exchanger networks", Chem. Eng. Sci., Vol.38(5), 745-763.

6. Linnhoff, B., et. al., 1982, "A user guide on process integration for the efficient use of energy", I Chem E, England.

7. Ranade, S. M. et. al., 1986a,
"Industrial heat pumps : Approximate sizing and placement using the grand composite",
Proceedings of the 8th IETC, Vol.1, Public Utility Commission of Texas, Houston, 194-203.

8. Ranade, S. M. et. al., 1986b,
"Industrial heat pumps : A novel approach to their placement, sizing and selection",
Proceedings of the 21st IECEC, Vol.1, Sandiego, California, 389-397.

9. Ranade, S. M., Hindmarsh, E. and Boland, D., 1987,
"Industrial heat pumps : New insights on their integration in total sites",
Proceedings of the 3rd International Symposium on Large Scale Applications of Heat Pumps, BHRA, Cranfield, U.K., 88-98.

10. Townsend, D. W. and Linnhoff, B., 1983,
"Heat and power networks in process design",
AIChE J,(29), 742-748.

Chapter 2

PROCESS INTEGRATION OF VARIOUS TYPES OF HEAT PUMP

P-Å Franck, E Wallin and T Berntsson (Chalmers University of Technology, Gothenburg, Sweden)

INTRODUCTION

To install a heat pump correctly in an industrial process is not only a question of heat pump optimization. The way the integration is made will at least have as much influence on the final result as the efficiency of the heat pump itself. Therefore a simultaneous optimization must be made which takes both the heat pump and the process into account.

Heat pump optimization is a difficult but rather well known subject mostly due to the existence of large computer codes which can calculate the performance of various designs and cycles at varying outer conditions. Process integration of the heat pump is, however, not that well known and will therefore be discussed in more detail.

The tool used here in process integration is the pinch technology. Pinch technology is a method by which complex industrial processes can be analysed and it has during the last decade played a major role in understanding complex industrial processes.

PROCESS INTEGRATION PRINCIPLES

To integrate a heat pump a good knowledge of the process is necessary. In this respect pinch technology is a very powerful tool in process analysis, see Linnhoff et al. (1982).
In a pinch technology analysis the process streams are characterized by their start and final temperatures and by their heat flow rate, i.e the mass flow multiplied by the heat capacity. Streams requiring cooling are called hot streams, whatever their absolute temperatures, and streams requiring heating are called cold streams.

Heat Pumps – Proceedings of the 4th International Conference, pp.19-26

One of the most important features of pinch technology is that it is possible to identify a temperature (the pinch) in a process which divides the stream system into two separate parts. In the part above the pinch there is a net heat deficit, and heat must be added to the system by a hot utility. If a cold utility is applied above the pinch it follows that the demand for the hot utility will increase by the same amount. Thus, valuable heat is just removed by cooling. On the other hand, in the part below the pinch there is an excess of heat that must be removed from the system by a cold utility. Any heat added below the pinch must also be removed. Hence, in a well designed process no cold utility should be used above the pinch and no hot utility below the pinch.

Composite curves are useful tools in pinch technology. These consist of two curves representing the composite hot and cold streams. The curves are constructed by adding the heat content of all hot and cold streams respectively in each temperature interval. The accumulated heat content can then be plotted on a temperature/heat content diagram, see Figure 1. It is possible to exchange heat from the hot streams to the cold streams where the two curves overlap horizontally, and the vertical distance between the curves represents the temperature difference in the heat exchange. Where the two curves do not overlap, external heating or cooling by plant utilities is necessary.

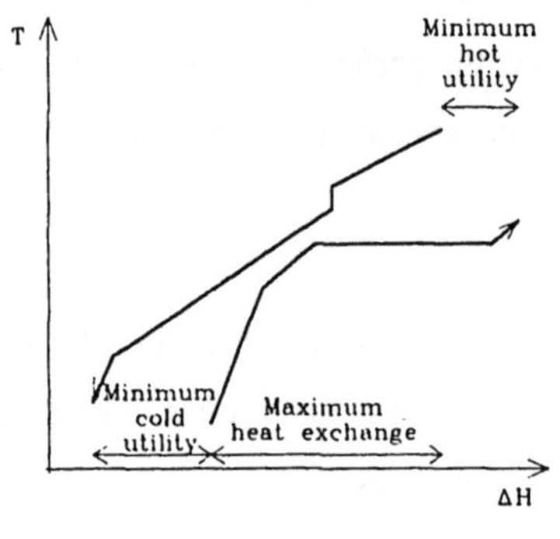

Figure 1. Composite curves.

From these facts it is obvious that a heat pump can be placed in three different locations: below, above, or across the pinch, see Berntsson (1987).

Placing the heat pump below the pinch does not result in any external utility saving because, if no hot utility is used below the pinch, heat is simply pumped around within that part of the system. But if, for some reason, a hot utility is applied below the pinch, a heat pump can either reduce its use or even replace it. The reduction is then equivalent to the prime energy consumption of the heat pump. This can, however, also be done - and probably more cheaply - by a better heat exchange system. Neither will a heat pump placed above the pinch save any external utility, although it may offer some advantages if hot utilities at different temperature levels are required. This can result in a reduction in the higher temperature utility requirement and a corresponding increase in the lower temperature utility demand. For this arrangement to be possible the temperatures must, of course, be such that the heat pump can operate between them. Although a heat pump placed below or above the pinch is thermodynamically possible, the economics of such cases are doubtful and no further consideration will be given to them in this paper. It should, however, be pointed out that an incorrectly placed heat pump, which does not save any energy, can be economic if the alternative is very expensive heat exchanging. The reason to the expensive heat exchanging can be a large distance between the streams, corrosive streams etc.

By placing the heat pump across the pinch, heat is extracted from a system where there is an excess of heat and delivered to a system where there is a heat deficit. In this way the demand for both the hot and the cold utilities is reduced. Thus a heat pump placed across the pinch is the correct way of integrating it into a process.

Above the pinch, heat should be delivered to the process from the heat pump. At each temperature level, the maximum amount of heat that can be delivered is limited to the horizontal distance between the composite curves at that temperature. The reason for this is that the hot streams must be warmer than the cold ones to make process heat exchange possible. If this distance becomes smaller at higher temperatures, this limits the possible heat transfer since a new pinch is then created. The introduction of the heat pump also means that the process heat exchanger area increases as a result of smaller driving forces in the heat exchangers. The reduction in driving forces is due to the fact that some of the cold streams now act as heat sinks to the heat pump. Consequently, the hot streams which formerly exchanged heat with these cold streams have to exchange heat with higher temperature cold streams.

Below the pinch similar conclusions can be drawn.

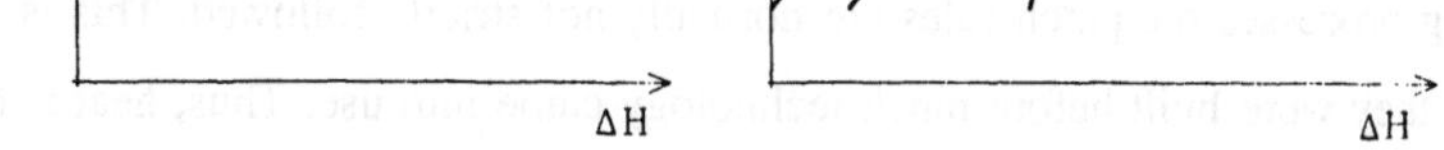

Figure 2. Tight and wide open composite curves

The principles discussed above are illustrated in Figure 2 in which it is assumed that heat is extracted from and delivered to all streams. In the figure two types of composite curves are shown, one tight and one wide open. The reduction of the driving forces is clearly seen. It can also be concluded that, for a given size of the heat pump, the process heat exchanger area increases when the source and sink temperatures come closer to each other. At the same time the heat pump cost decreases which leads to an optimization problem. The main parameters to be optimized are the size of the heat pump and the inlet temperatures to the heat pump of the heat source and demand. One obvious solution is that the size of the heat pump is zero, i.e. there should not be any heat pump at all in the process.

To solve this optimization a large computer program has been developed. By the aid of a mathematical library and cost data the the program finds the heat pump placement in a process which has the lowest annual cost. The parameters taken into account in this optimization are: reduction in utility costs, operation and investment costs of the heat pump, and investment costs of the process heat exchangers. To be able to do this calculation the composite curves of the process must be known. Furthermore, the type of heat pump must be specified. Characteristics for the most common heat pump types are today available in separate modules, and new types can easily be added.

ENERGY SAVINGS BY INCORRECT PLACED HEAT PUMPS

The most common way to install a heat pump in a process has been in connection to a unit operation such as a distillation column or an evaporator. Another common arrangement is to upgrade waste heat of various temperatures to a circulating hot water system or to a low pressure steam system. From basic pinch technology knowledge it is obvious that these arrangements will not necessarily achieve any energy savings. Nevertheless big energy savings are reported from most installations although the pay back period has not always been within acceptable limits. There are two reasons to the energy savings:

i: In the unit operation oriented integration approach no energy savings are achieved if the pinch temperature does not lie within the temperature limits of the operation. If the unit operation is the major energy consumer in the process the pinch will normally lie within the operation. In such cases the heat pump is correctly placed and the desired savings are obtained. If the unit operation is not the major energy consumer, which can be the case in more complex processes, the energy consumption increases (the heat pump is then placed below the pinch) or the original utility is replaced by the heat pump drive energy (this corresponds to the case when the heat pump is placed above the pinch). Thus, totally no energy savings are obtained but the energy consumption of the unit operation decreases.

ii: In existing processes the pinch rules are normally not strictly followed. This is in most cases due to the fact that they were built before pinch technology came into use. Thus, heat is transferred across the pinch and coolers can even be placed above the pinch. Both these heat flows violate pinch technology rules and result in an increased energy consumption. If a heat pump is installed in such a process in a way that these heat flows are utilized as heat sources, then the energy consumption will decrease by the same amount. This is due to the fact that unwanted heat flows from the system above the pinch where there is a deficit of heat are reduced. Thus a heat pump placed above the pinch can save energy if the pinch rules are not followed in the process.

SELECTION OF HEAT PUMP TYPE

For the process industry there are today three types of heat pumps on the market that are of major interest: closed cycle compression, mechanical vapour recompression and heat transformer. Other types such as the combined compression-resorption cycle and the compression cycle with non-azeotropic mixtures have not yet reached a commercial break through. In future, however, they will be an interesting alternative because of the restrictions that are put on the use of CFC's in ordinary compression heat pumps and the high temperatures at which they can operate.
The three types of highest interest have all different characteristics and demands and will therefore be more or less suitable in a specified application. As a first step in the selection of type the pinch temperature and the shape of the composite curves can serve as guidelines.

The **electrically driven compression heat pump** should operate with a COP of at least 3 but preferably higher. For this to be possible the temperature lift cannot be too high, which implies that the composite curves must open up not far away from the pinch temperature. To make full use of the heat pump the available source should also be of the same size as the sink. One condensing stream below and one evaporating one above the pinch are therefore ideal conditions. The fact that the CFC's used today cannot work at condensing temperatures above 120 °C reduces the potential for compression heat pumps to processes with a pinch temperature of 120 °C and below.

The **mechanical vapour recompression heat pump** (MVR) has the advantage to work at higher condensing temperatures than 120 °C. For this type the limitation lies instead at the other operating end, since evaporating temperatures below 80 °C are normally not of economical interest. Thus, MVR is suitable for processes with a pinch temperature above 100 °C. Because of the relatively small temperature lift per stage the composite curves need to open up within this temperature interval.

The **heat transformer** operates practically without electricity and is capable of upgrading half of the heat source to a maximum temperature of 150 °C. Since the largest temperature lift is about 40 °C to 50 °C, the pinch temperature should be between 60 °C and 120 °C to fit the heat transformer. The best composite curves have a heat demand which is not larger than half of the heat source and consist of condensing and evaporating streams, respectively. Process integration of heat transformers is described in detail by Berntsson et al. (1989).

METHOD OF PROCESS INTEGRATION OPTIMIZATION AND ASSESSMENT

From above it is clear that a heat pump operating across the pinch will save both hot and cold utility. However, the annual cost, which is the sum of the annual operation and capital costs, will not necessarily be lower in a process with a heat pump compared to the same system if only heat exchange is used. This is due to the capital costs of the heat pump itself and to the larger heat exchanger area needed. This leads to a situation where there is a competition between heat exchanging and heat pumping.

For a system which only uses heat exchanging the optimal solution can be found from the trade off between heat exchanger area and energy costs. By shifting the composite curves apart the driving force for the heat exchange increases, leading to a lower capital cost but at the same time to a higher operation cost since the energy consumption increases. With known cost data this optimization is fairly straight forward and accepted. The distance between the composite curves (ΔT_{min}) is obtained as a result from the optimization.

For a system which uses both heat exchanging and heat pumping the optimization is more complex. First of all it is the same trade off between heat exchanger area and energy as in the pure heat exchange case. Secondly it is the trade off between energy savings due to the heat pump and extra heat exchanger area needed. The results that come out of such an optimization are: heat pump size, placement, and ΔT_{min}. This optimization must be made for those types of heat pumps that might be of interest in that specific situation.

When the system which only uses heat exchanging and the system which uses both heat exchanging and heat pumping are optimized, the annual cost of them can be compared and the potential for a heat pump installation can be revealed.

EXAMPLES

The optimization procedure described above is here shown for two different processes, one pulp industry (case I) and one organic process (case II). The processes are represented by the composite curves in Figure 3 and 4.

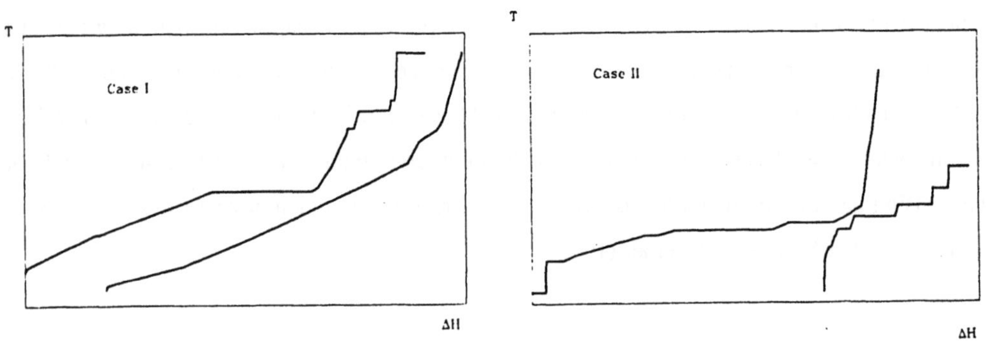

Figure 3 and 4. Composite curves for case I and II.

The hot pinch temperature is 62 °C in case I and 118 °C in case II. Thus the pinch temperatures represent the operation limits for the heat transformer and the upper limit for a compression heat pump. Therefore, as an example, the optimization is performed for these two types of heat pumps. Furthermore the optimization is limited to the existing ΔT_{min}.

Assumptions

The COP of the electrically driven compression heat pump varies with temperature lift, absolute temperature, working fluid, and compression type. In the optimization an average behavior of a large scale heat pump with working fluids used today is assumed. The resulting Carnot efficiency varies therefore between 0.50 and 0.63 depending on the temperature lift.

The technical behavior of the heat transformer is obtained with the aid of a computer program by Holmberg (1988). With a specified COP (in this case 0.48) the program calculates the necessary heat transfer area in the heat transformer.

In calculating the process heat exchanger area pure countercurrent heat exchange is assumed.

The cost data used is shown in Table 1.

Table 1 Cost data

Fossil fuel cost	32.8 US$/MWh
Heat exchanger cost	32 970 + 656 · area US$
Compression heat pump cost	245 US$/kW
Heat transformer cost	460 to 760 US$/kW (depending on operating temperatures)
Annuity factor	0.20
Annual operating time	8000 h

Results

In Table 2 the results are summarized. In the table the energy savings due to the heat pumps and the pay back period are shown for two ratios between the electricity price and fuel price (γ-value). Furthermore, the temperatures between which the heat pump operates are shown.

Table 2 Results

		T_{source} °C	T_{sink} °C	Energy Savings %	PBP years
Case I:					
Heat transformer		59	68	86	2.1
Compr. heat pump	γ=2	58	70	86	1.3
	γ=4	58	70	86	1.7

Case II:

Heat transformer		84	120	38	1.5
Compr. heat pump	γ=2	97	109	39	1.0
	γ=4	97	109	39	1.3

From the table it can be concluded that the pay back periods are short and acceptable and little dependent on the γ-value, since the COP of the compression heat pump is very high. It can also be seen that the temperatures between which the heat pumps operate are nearly the same in all cases. This is not surprising since the placement of the heat pump in these cases is fairly straight forward due to the shape of the composite curves. More surprising is that the energy savings are the same in all cases. The reason for this is that a new pinch is created above the pinch and consequently any larger heat pump cannot be installed. This situation also leads to the consequence that the two types of heat pump utilizes very different amounts of heat source. Thus, the integration below the pinch will be quite different.

CONCLUSIONS

The way the process integration of a heat pump is made will play a major role in the energy savings achieved. To integrate a heat pump without taking pinch technology into account will not necessarily save any energy.

The selection of a suitable heat pump type for an application must be based on the pinch temperature and the shape of the composite curves. With this information available it is possible to find the installation which has the lowest annual cost. The pay back periods are then often low.

REFERENCES

Berntsson K. et al. 1989. Heat Transformers in Industrial Processes. CADDET Analyses Series No. 2, 1989.

Berntsson T. (1987). Future Prospects for Industrial Heat Pumps in Europe. Proceedings of the 1987 IEA Heat Pump Conference, Orlando USA, pp 303-324.

Holmberg P. 1988. System Studies and Optimization of the Single-Stage Absorption Heat Transformer Cycle. Licentiate thesis, Department of Heat and Power Technology, Chalmers University of Technology, Sweden.

Linnhoff B. et al. 1982. A Users Guide on Process Integration for the Efficient Use of Energy. Published by the I. Chem. E., Rugby U.K.

Chapter 3

EXPERIMENTAL HYBRID HEAT PUMP OF 1000kW HEATING CAPACITY

Gy Bergmann and G Hivessy (Institute for Energetics, Budapest, Hungary)

ABSTRACT

The hybrid heat pump is a compression heat pump the working fluid of which is a non-azeotropic mixture. From such cycle - in comparison with the conventional heat pumps - the following advantages are expected:

- higher COP;
- lower pressure ratio;
- lower absolute pressure;
- higher temperature gap can be bridged over in one stage;
- wider area of application by setting different mixture concentration.

In a poultry slaughterhouse, for example, the following task can be accomplished by a heat pump:

Heat sink (wash water)	15/85 °C
Heat source (cooling water of condensers)	25/5 °C

Due to the significant temperature changes (especially that of the heat sink) and to the large temperature gap, it is the hybrid heat pump the application of which seems to be expedient. In order to solve the task, hybrid heat pump cycles operating on ammonia-water mixture were examined. The most advantageous connection was selected and a pilot plant was built accordingly. This connection is characterized by the great extent of the internal heat exchange and by wet compression. Both factors improve the COP efficiently, and the internal heat exchange reduces the pressure ratio, as well.

The paper describes the most important technical data of the heat pump. The test operation of the experimental equipment was started in August 1989 and is expected to be terminated in June 1990. Thereafter the results of the measurements will be evaluated.

1. THE HYBRID HEAT PUMP

The hybrid heat pump is a compression heat pump the working fluid of which is a non-azeotropic mixture. Such media evaporate and condense at variable temperatures (while the pressure is constant). This can be explained as follows.

In the course of the heating process in the evaporator, the more volatile component is expelled from the solution, while the less volatile component evaporates, i.e. not only evaporation, but also

Heat Pumps – Proceedings of the 4th International Conference, pp.27-40

desorption takes place, resulting in continuous change of the composition of the remained solution. Similarly, in the course of the cooling process in the condenser, condensation as well as absorption take place. The most common working fluid of that kind is the aqueous solution of ammonia. All known experimental plants of industrial size operate on this mixture. It is the gliding temperature of evaporation and condensation that makes cycles with such fluids well applicable in cases where the temperatures of the heat source and the heat sink change considerably in the course of heat dissipation and heat absorption.

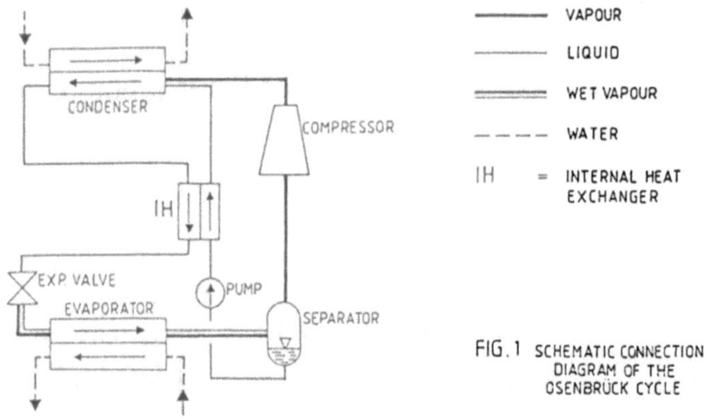

FIG. 1 SCHEMATIC CONNECTION DIAGRAM OF THE OSENBRÜCK CYCLE

Theoretical investigations of the hybrid heat pump did take place already several decades ago [2, 3, 4], though with other denominations (e.g. compression machine with solution cycle). At that time, however, no heat pump of this type was actually realized.

About 1980 the circumstances became ripe for the theory being put into practice, therefore the hybrid heat pump got into the highlight again. Several reports were published describing theoretical and practical investigations, as well as surveying the status of research and development work, and plants of industrial size were established [10, 13, 14, 17, 21, 23, 24].

The expected advantages of the hybrid heat pump compared to conventional heat pumps are as follows [22]:

In case of external media with variable temperatures:

- higher COP;
- lower pressure ratio;
- higher temperature gap can be bridged over in one stage.

General benefits:

- advantageous pressure ranges within wide temperature limits, reached by setting the adequate composition;
- lower absolute system pressures in most cases;
- the mixtures (e.g. ammonia-water) are able to replace CFCs in the entire range of application.

The hybrid heat pump, therefore, has all advantages of heat pumps as far as energy conservation is concerned, but compared to the conventional heat pumps it has some additional advantages as well, enabling the economic extension of heat pumping towards industrial applications demanding higher temperatures (e.g. over 100 °C). The utilization factor of most industrial consumers is much higher than that of the most common space heating applications, because their heat demand is much less seasonal, making this way the heat pumping more economical. The hybrid heat pump is more advantageous from the aspect of environmental protection as well.

The theoretical principles of the hybrid heat pump can be found in details in an article published in 1989 [20].

2. SELECTION OF THE PILOT PLANT CONNECTION

Some years ago, the Institute for Energetics (EGI), Budapest, decided to erect an experimental heat pump of industrial size, with 1000 kW heating capacity. The project was supported by the government in the framework of the National Middle-Term Research and Development Plan (OKKFT G/4 program).

The experimental equipment was erected at a poultry slaughterhouse, where hot water supply of 82 °C had to be ensured by warming up tap water.

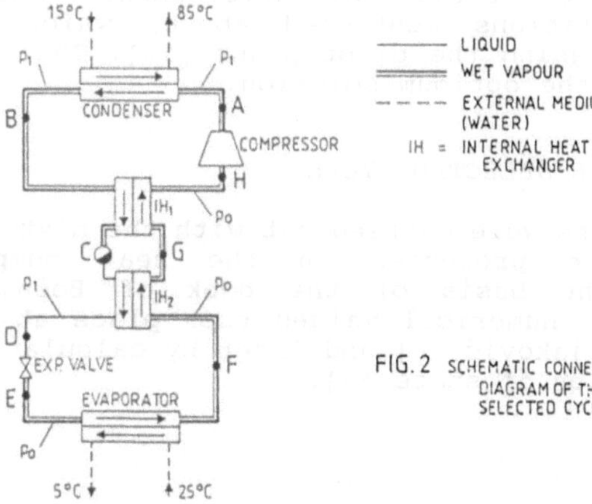

FIG. 2 SCHEMATIC CONNECTION DIAGRAM OF THE SELECTED CYCLE

As heat source the warmed-up cooling water of the condensers of the refrigerating machines was used. The design parameters of the heat pump were determined as follows:

Heat source (cooling water):	25/5 °C
Heat sink (wash water):	15/85 °C

The big temperature gap to be bridged over (80 °C), the high temperature of the heat sink (85 °C), and last but not least the significant temperature change of the external media, especially that of the heat sink (70 °C), make the task expressly suitable for emphasizing the advantages of hybrid heat pumping.

Thorough theoretical investigations [13, 15, 18, 19] and laboratory examination of processes in the heat exchangers [16] preceded the establishment of the pilot plant. As far as the connection is concerned, in the end six characteristic variations were compared [18].

The earliest known reference to hybrid heat pumps is the Osenbrück patent from 1895 [1], the connection diagram of which is shown in Fig. 1. This is not the simplest possible hybrid heat pump, since it contains, in addition to the absolutely necessary elements (evaporator, condenser, pressurizer, pressure reducer), also an internal heat exchanger. In the evaporator and condenser of the hybrid heat pump the temperature of the working fluid changes during the heat transfer process, so in this respect the hybrid cycle is similar to the heat pumps with gaseous working media. Therefore the internal heat exchange can reduce the pressure ratio at the hybrid heat pump as well.

In the course of further development, as far as the connection of the hybrid heat pump is concerned, two main aspects seemed to be the most important. One is wet compression, the first reference to which we discovered in the invention of Najork [7]. The importance of wet compression can be explained by the fact that as a result of continuous recooling of the compressed vapor it reduces the compression work and the final temperature of the compression. Najork modified the Osenbrück cycle by conducting the liquid warmed up in the internal heat exchanger into the compressor instead of the condenser.

Another development trend is to increase the internal heat exchange. The first step to technical solution of that kind is the cycle [8], where the wet vapor getting out from the evaporator is not separated, the entire low-pressure working fluid is let into the internal heat exchanger and from there into the compressor, taking the advantage of wet compression as well. This cycle was improved by a further step [12] which renders possible further increase of internal heat exchange by conducting wet vapor instead of liquid from the condenser into the internal heat exchanger. The connection diagram of this variation is indicated in Fig. 2.

After the investigation of the above development trends we defined the six connection variations mentioned above, which were thoroughly examined before designing the pilot plant [18]. The connection as per Fig. 2 proved to be the optimum solution.

3. DESCRIPTION OF THE SELECTED CYCLE

The cycle calculations were carried out with the mixture of 80% ammonia and 20% water. The processes in the heat pump were examined theoretically on the basis of the book of Bošnjaković [5], the determination of the numerical values took place at the beginning by the diagrams of Bošnjaković [6] and later by calculations based on the Trepp-Ziegler equation of state [9].

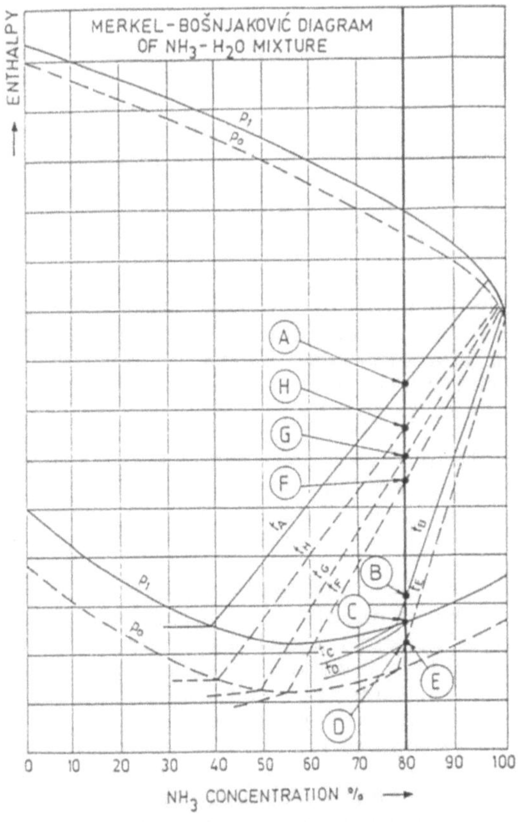

FIG.3 ENTHALPY VS. CONCENTRATION OF NH_3-H_2O MIXTURE
HYBRID HEAT PUMP PROCESS
IN ENTHALPY - CONCENTRATION DIAGRAM

Figure 3 shows the processes accomplished by the connection as per Fig. 2 in the enthalpy-concentration diagram. The working fluid of state A (wet vapor) enters the condenser at pressure p_1, where at continuously decreasing temperature the greater part of the working fluid gets into liquid state, while the dissipated heat is transferred to the washing water warming up. Hereupon, the wet vapor in state B enters the first part of the internal heat exchanger (vapor-cooling internal heat exchanger), where, at further decreasing temperature, the entire working fluid gets into liquid state (C). Then the working fluid enters an intermediate tank, from which the saturated liquid flows into the second part of the internal heat exchanger (liquid- cooling internal heat exchanger). In the latter the working fluid is subcooled and gets into state D. From there the liquid-state working fluid flows through the expansion valve, where its pressure drops to p_0, while its temperature decreases considerably, and a significant part of it gets into vapor state.

The working fluid of state E enters the evaporator, where, at pressure p_0 and at continuously increasing temperature, it absorbs heat from the cooling water being the heat source. In the course of this process significant part of the working fluid gets into vapor state (F). Then the working fluid absorbs, at continuously increasing temperature and vapor content, the heat released by the high-pressure working fluid in both parts of the internal heat exchanger (points G and H). Thereafter the compressor increases the pressure of the working fluid again to p_1, which raises the temperature of the working fluid considerably and its vapor content in a smaller degree.

Figure 3 shows the enthalpy-concentration diagram of the ammonia-water mixture, indicating the boundary curves for the pressures p_0 and p_1 and the characteristic points of the cycle with their isotherms. Therefore the proportion of the two phases (liquid and vapor) and their concentration in the given points can be followed in the diagram. It can also be perceived that in this diagram every point of the cycle falls in one single vertical straight line. This is due to the fact that it is the entire working fluid that flows in every section of the cycle, i.e. the two phases are not separated, and this way the concentration of the working fluid is equal everywhere.

This renders possible to present the cycle in a relatively simple way in the temperature-entropy diagram as per Fig. 4. This way of presentation is favorable, among others, also by showing the irreversibilities of the cycle, e.g. it illustrates the temperature changes of the working fluid in the heat exchangers, so the temperature gaps in the heat-exchangers are easier to survey.

4. INTERNAL HEAT EXCHANGE

The temperature-entropy diagram makes easily comprehensible - due to the reasons mentioned - also the importance of the internal heat exchange. With a compression heat pump, internal heat exchange means that a part of the heat transferred by mechanical work to higher temperature level is reduced to lower temperature. This, as a matter of fact, causes a loss which is the less the less is the temperature difference of the fluid at the two sides of the heat exchanger. Theoretically the loss can be eliminated if the heat exchanger surface is infinitely great and the temperature graphs of the fluids flowing on either side are identical. Practically such losses always occur; in addition, the heat exchanger costs money, and the pressure loss of the working fluid in the heat exchanger has to be covered by mechanical work.

That is why generally no internal heat exchange is applied with compression heat pumps operating on vaporous working media. Though it occurs occasionally, in most cases it is only used for superheating the medium leaving the evaporator, in order to avoid the liquid incidentally remaining in the working fluid.

The situation is quite different with gaseous working fluids. In that case internal heat exchange is generally applied, in spite of its disadvantageous impacts. The reason for this is hidden in the fact that internal heat exchange has special advantages connected with the temperature graphs of the working fluid in the course of heat exchange with external media. With such cycles, as a matter of fact, both heat input from the heat source and heat output to the heat sink take place with significant change of the temperature of the working fluid, since in that case the phase of the working fluid does not change. As a consequence of this, internal heat exchange in many times renders possible significant reduction of the pressure ratio. This has the following well-known advantages:

- higher compressor efficiency;
- less stages of compression in many cases (e.g. one compressor instead of two);
- lower absolute pressure (lower nominal pressure, therefore less expensive equipment);
- the losses connected with the efficiency of the compression and expansion decrease.

With the hybrid heat pump, the working fluid undergoes a change of phase in the course of heat dissipation or heat absorption, but in the meantime its temperature changes as well, as e.g. it is shown by the diagram in Fig. 4. For this reason, the role of internal heat exchange is similar to that of heat pumps operating on gaseous working fluids. The only difference is, that the working fluid of the hybrid heat pump is vaporous, therefore, instead of the expansion turbine, expansion valve is used generally. In the latter case, the saving obtained by decreasing the pressure ratio is still higher.

Based on the above, it is obvious that with the hybrid heat pump the significance of internal heat exchange is similar to that of gaseous cycles, or even higher. That is why internal heat exchangers are applied in most cases. This was already contained in the Osenbrück invention (Fig. 1), although it was a liquid-to-liquid heat exchanger with relatively low capacity. As already mentioned, internal heat exchange was increased step by step in the course of development of hybrid heat pumps.

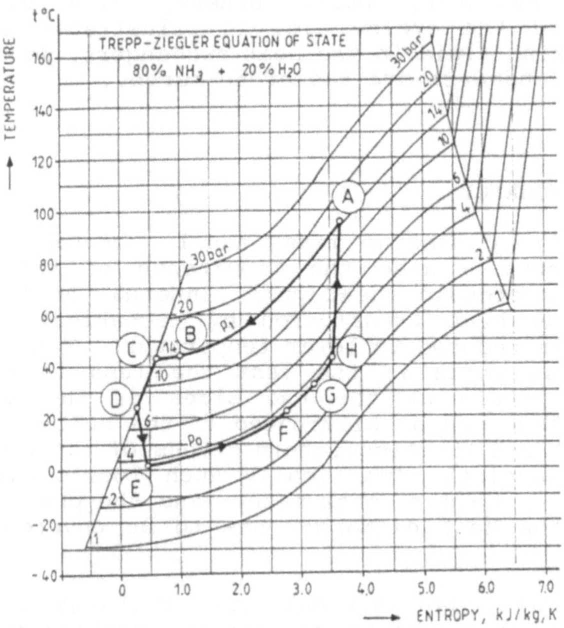

FIG.4 HYBRID HEAT PUMP PROCESS IN TEMPERATURE – ENTROPY DIAGRAM

Figure 5 presents the importance of the internal heat exchange, as far as the pressure ratio is concerned, in case of cycles fitted to the mentioned application in a slaughterhouse. The figure indicates three cycles between the same temperature limits (t_0 = 2 °C and t_1 = 98 °C).

When making the calculations, it was assumed that in all three cases compressor efficiency was equal to 67 %. The COP is understood here as the ratio of the condenser heating output and the compressor shaft power.

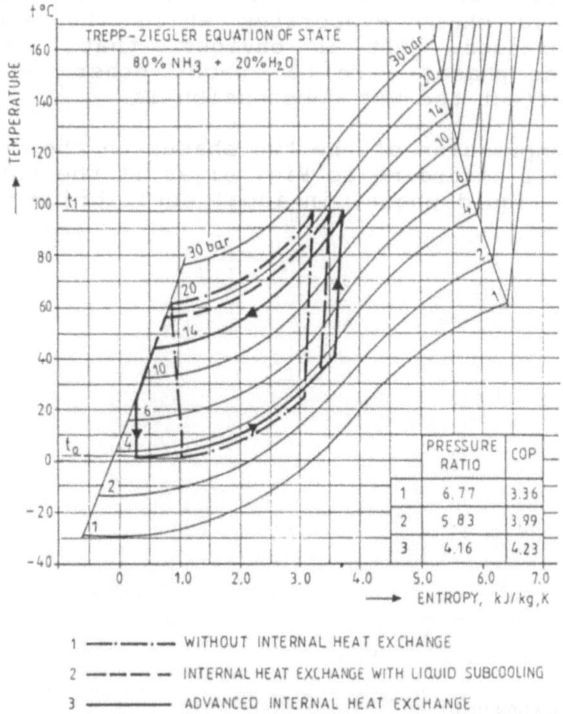

FIG. 5 SIGNIFICANCE OF THE INTERNAL HEAT EXCHANGE

5. WET COMPRESSION

It is well-known that in case of compression of a gas between given pressure limits, the intermediate recooling decreases the compression work as well as the final temperature of the medium. Recooling takes place generally between the stages of compression. Another well-known solution is when liquid (e.g. water with air compressors) is injected into the compressor. The injected liquid, evaporating, recools the warming working fluid continuously, which is similar to intermediate recooling to be accomplished in an infinite number of stages.

Also with vaporous working fluids the idea may come up that, by allowing liquid to enter the compressor, the favorable effects mentioned above could be obtained. One of the possibilities is when this liquid is the liquid phase of the working fluid. It can be injected separately into the compressor, or it can enter the compressor as a part of the wet-vapor-state working fluid. In this case wet compression takes place, and for that purpose a suitable compressor should be applied.

In consideration of the above respects we decided to operate the hybrid heat pump pilot plant with wet compression. Wet compression - as already mentioned - theoretically implies recooling in an infinite number of stages and this way significantly reduces the power demand of the cycle. This advantageous effect, however, prevails only to such extent by which the liquid in the compressor can follow the change of state of the vapor. In the course of compression the volume of the vapor phase decreases, therefore it warms up, while the temperature of the liquid phase changes only scarcely. The much warmer vapor phase heats up the liquid, but the latter does not reach the state of equilibrium with the vapor phase till the end of compression.

This means that on leaving the compressor, the working fluid is a non-equilibrium mixture which consists of overheated vapor as well as liquid drops at the same time. The two phases of the working fluid, flowing in the discharge conduit of the compressor, reach the state of equilibrium and this way the identical temperature after passing a certain section of the conduit. In the mentioned section one part of the liquid phase evaporates while the temperature of the vapor phase decreases. It is obvious, that the closer the temperatures of the phases are to each other at the compressor outlet the more perfect the wet compression is.

The real common compression of the two phases in the compressor can be simulated by a model according to Fig. 6: as if the entire vapor phase and a part of the liquid phase, in ideal equilibrium with each other,

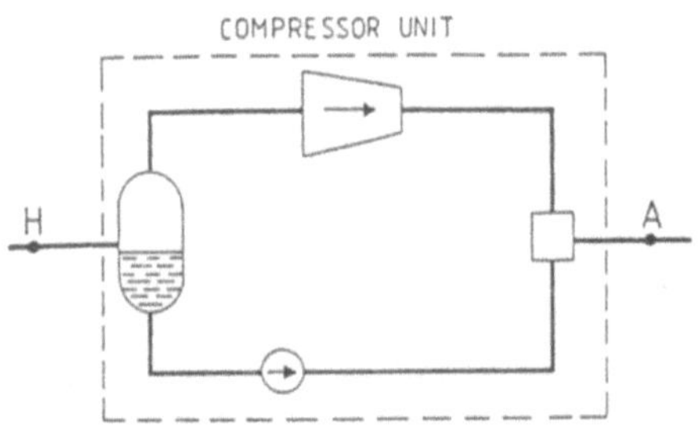

FIG. 6 CALCULATION MODEL FOR WET COMPRESSION

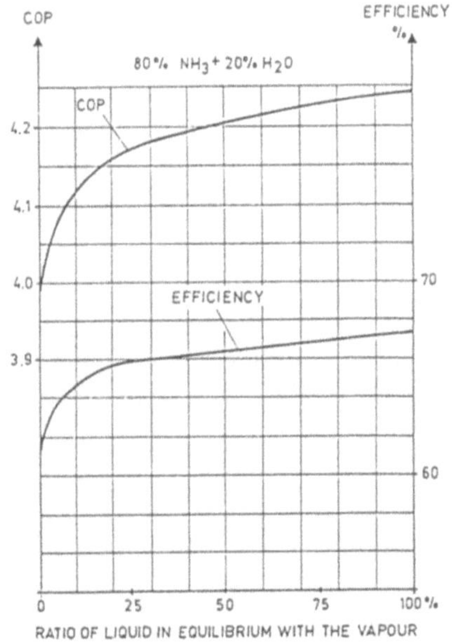

FIG.7 COP AND EFFICIENCY OF COMPRESSION VS. EQUILIBRIUM RATIO DURING WET COMPRESSION

were pressurized by the compressor, the other part of the liquid phase, however, were delivered by a pump. In case of perfect wet compression the pump of the model does not deliver anything. On the contrary, if between the vapor and the liquid in the compressor no heat is transferred at all, then the compression can be considered separated, i.e. all liquid is delivered by the pump. In general, according to the model, the degree of the disequilibrium of the process can be characterized by the part of the liquid phase delivered by the pump.

With this model, calculations were made in order to determine the effect of the disequilibrium, using technical data of a planned running condition of the pilot plant. In the course of the calculations, it was assumed that the internal efficiency of the compressor was 67% and the COP of the cycle with perfect wet compression (if the pump did not deliver anything) was 4.23. Figure 7 indicates the results of the calculation, i.e. the COP against the part of the liquid phase, being in equilibrium with the vapor phase. It can be perceived that in case of perfect wet compression the COP is 4.23, while it is only 3.99 in case of entirely separated compression. The figure also indicates the internal efficiency of the compression. This, in case of perfect wet compression, is equal to the internal efficiency of the compressor and decreases with the increase of liquid delivery by the pump. An important result is that deviation from the perfect wet compression to a certain limit causes only a slight loss.

6. THE PILOT PLANT

Some information on the pilot plant were published earlier [22]. Fig. 8 shows the schematic connection diagram of the heat pump pilot plant. In this respect the following details are worth mentioning:
The heat exchangers are designed in a way to approach the perfect counterflow, which is absolutely necessary in order to utilize the gliding temperature graph. The heat exchangers were designed on the basis of thorough theoretical investigations and laboratory tests [16]. The wet-vapor-state working fluid flows in the tubes of the shell-and-tube heat exchangers. To keep up the approximate equilibrium of the two phases special mixing inserts are provided [11].

The oil-free screw compressor 1/CO suited for wet compression is the essential unit of the equipment. This is a specially manufactured Atlas Copco air compressor adapted for this purpose.

In order to make wet compression as perfect as possible, the theoretical connection as per Fig. 2 was completed with some units, shown in Fig. 8. Wet vapor in state H leaves the internal heat exchanger 3/IH-1 towards the compressor. Then liquid is separated from the wet vapor in the separator 9/LS. The vapor phase is sucked by the compressor, while the liquid phase is delivered by the pump 8/LP. The liquid can be injected into the vapor flow both before and after the compressor. The distribution of the liquid between the two branches can be varied, this way the most advantageous proportion can be set by experimental procedure.

With the purpose to intensify the heat transfer between the two phases, the injection process is fulfilled by atomizer nozzles, which transform the liquid into small droplets. This is especially important before the compressor. Therefore there are 30 atomizers (10/AN) on the suction side and only 3 on the discharge side (11/AN).

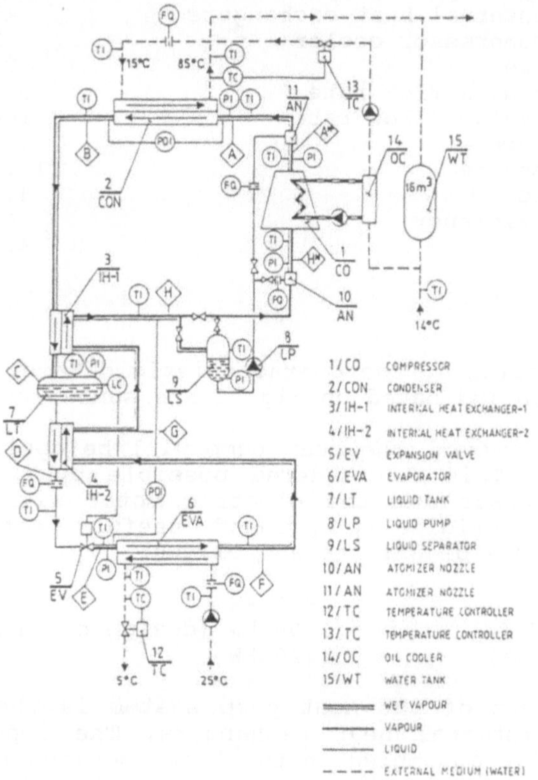

FIG. 8 HYBRID HEAT PUMP PILOT PLANT
CONNECTION DIAGRAM

The entire by-pass of the liquid separator is rendered possible with this equipment, in that case the whole working fluid is delivered in wet-vapor state by the compressor.

Another difference as compared to Fig. 2 is, that the oil, cooling and lubricating the compressor bearings and gear drive, is recooled by the oil cooler 14/OC, where the heat dissipated from the oil is also utilized for water heating. Tap water is preheated in the oil cooler before it enters the condenser 2/CON.

The warm water tank 15/WT is connected to the heat sink. It is destined, on one hand, to compensate the discrepancy between the changing heat demand of the consumer and the constant heat output of the heat pump, and, on the other hand, to ensure steady-state operation required for experimental measurements.

In addition to the mentioned completion of the connection, there are also other differences between the realized equipment and the variant which proved to be the optimum solution in the first phase of preparation. This earlier variant (connection diagram as per Fig. 2, working fluid cycle as per Fig. 3 and 4) was further examined later. As a result of this, increased ammonia concentration turned out even more favorable. The final design characteristics are indicated in Table 1.

Design parameters of the pilot plant — Table 1

Heat source (water)	25/5 °C
Heat sink (water)	15/85 °C
Working fluid	85% NH_3 + 15% H_2O
Condenser capacity	1000 kW
Evaporator capacity	770 kW
Capacity of internal heat exchangers	170 kW
Capacity of compressor cooler	14 kW
Compressor power	230 kW
Working fluid mass flow rate	0.97 kg/s
Suction side volume flow rate	1088 m^3/h
Suction pressure	3.7 bar
Discharge pressure	15.2 bar
Pressure ratio	4.1
Compressor efficiency	67%
Internal COP	4.35

In this design condition, the characteristic data of the working fluid cycle in the points indicated on Fig. 2 are summarized in Table 2.

According to the program, the heat pump will be operated and tested at different outputs. This is rendered possible by a V-belt gear drive between the compressor and the electric motor with variable rate of transmission. The originally planned useful heating outputs with ammonia-water mixture of 85% are as follows:

transmission rate I.	820 kW
transmission rate II.	1000 kW (design condition)
transmission rate III.	1220 kW

An important element of the heat pump system is the liquid tank 7/LT between the two internal heat exchangers. The impulse of the high-pressure float valve assembled on the tank operates the expansion valve 5/EV and this way controls the working fluid flow through the valve. By maintaining a constant level in the tank, it is assured that always saturated liquid leaves the internal heat exchanger 3/IH-1.

Characteristic points of the hybrid heat pump cycle Table 2

	A	B	C	D	E	F	G	H
V [m^3/h]	321	36.6	5.39	5.17	87	864	964	1088
p [bar]	15.1	13.9	13.9	13.9	3.9	3.7	3.7	3.6
t [°C]	104	42	41	22	2	21	30	41
i [kJ/kg]	1283	195	89	-4	-4	827	920	1027
x_L[%]	17.9	90.5	100	100	92.6	33.3	28.7	24.1
x_V[%]	82.1	9.5	-	-	7.4	66.7	71.3	75.9
ξ_L[%]	38.4	83.4	85	85	83.8	56.0	49.5	42.1
ξ_V[%]	95.4	99.9	-	-	99.9	99.8	99.6	99.0

Explanation of signs:

- V - working fluid volume flow rate
- p - pressure
- t - temperature
- i - enthalpy
- x_L - liquid phase ratio
- x_V - vapor phase ratio
- ξ_L - NH_3 concentration in liquid
- ξ_V - NH_3 concentration in vapor

Beyond that the control system of the heat pump is in connection only with the external media. At the heat sink the temperature of the water leaving the condenser 2/CON is set by the temperature controller 13/TC regulating the wash water flow rate. As far as the heat source is concerned, temperature of the cooling water leaving the evaporator 6/EVA at 5 °C is kept constant by the temperature controller 12/TC regulating the cooling water flow rate.

Another characteristic feature of the heat pump pilot plant is that the equipment is installed on a crowded industrial area. Therefore the heat exchangers were placed outdoors, while the compressor in the machine compartment at a distance of some 20 m from the heat exchangers. Thus, the different elements of the heat pump had to be linked with long pipelines of intricate traces, carrying wet vapor during operation. Therefore the questions arose whether the steady-state operation of the heat pump could be ensured among such conditions (would the liquid cause oscillations or not), and whether the liquid accumulated at various points of the pipeline could threaten the safety of operation during start-ups.

7. MEASURING AND EVALUATING SYSTEM

A measuring and evaluating system was established at the pilot plant. Figure 8 indicates the measured points of the working fluid and of the external media. These measured characteristic points are able to represent the change of state of the media.

At the points indicated and in the electric switchboard the following parameters can be measured:

- temperature
- pressure
- pressure difference
- fluid flow rate
- electric power

The last value, the measurement of which is not shown in Fig. 8, is one of the most important data of the heat pump, since it indicates the quantity of electric energy required for heat supply.

The measurements can be divided into two groups according to their tasks. The first one serves for the energetic evaluation of the heat pump. In this context the following values have to be determined:

- Useful heat output

It is obtained by measuring the flow rate, as well as the inlet and outlet temperatures of the washing water being the heat sink.

- Heat absorbed from heat source
It is obtained by measuring the flow rate as well as the inlet and outlet temperatures of the cooling water serving as heat source.

- Compression work
It can be determined by the electric energy input of the driving motors of the compressor and the liquid pump.

With the above data the COP of the hybrid heat pump at given temperatures of the external media can be determined.

The mentioned measurements concern the evaluation of the heat pump as a whole. The pilot plant has also the purpose to test the operation of the various components of the cycle, to check the accuracy of the design procedures, to draw conclusions from occasional discrepancies between measured properties of the working fluid in different points of the cycle and those calculated with the equilibrium equations of state. The second group of measurements complies with this requirement. The measurements at the points of the cycle marked by numbers in Fig. 8 provide a large amount of information. In the following only the most important aspects are summarized - instead of a detailed explanation.

- Concentration measurement
Determination of the composition of the working fluid circulating in the cycle has decisive importance with regard to the calculation of properties. This composition is not inevitably identical to the composition of the working fluid filled into the system, since liquid with ammonia concentration differing from the concentration of the whole working fluid can be separated at some places of the system consisting of vessels and extended piping, therefore the composition of the remaining part of the working fluid will also differ from that of the entire working fluid. The most suitable place for measuring is point C where there is saturated liquid being in contact with its equilibrium vapor. In this case pressure and temperature determine the concentration.

- Measurement of the mass flow rate
For that purpose point D is suitable since there the whole working fluid is in subcooled-liquid state, its pressure and temperature determine its density unambiguously (supposing the concentration is known).

- Heat transfer and fluid mechanics in heat exchangers
In most parts of the heat exchangers wet-vapor-state working fluid flows. The purpose of the measurements is to determine the heat transfer coefficients and pressure drops in the heat exchangers. Before the condenser in point A the pressure and temperature of the working fluid are measured, while in point B after the condenser the temperature, as well as the pressure drop between the points A and B. In case the mass flow and composition of the working fluid are known, the pressure drop indicates the resistance of the condenser in the given flow conditions, and on the basis of the pressure and temperature values the temperature graph of the condenser on working fluid side can be determined. Since that of the water side is already known, the overall heat transfer coefficient can be calculated.

The procedure is similar concerning the evaporator with the only difference that the two internal heat exchangers (of similar structure as the evaporator) have been combined - as far as evaluation is concerned - with the evaporator itself. I.e. in this case the change of the state of the working fluid between points E and H is examined. This is completed by the temperature measurement in point F.

- Wet compression
The entire compression process takes place between the points H and A, but the process is also examined in details. In the tank 9/LS there is saturated liquid in contact with its own vapor. Thus, the concentration of the separated liquid can be determined by measuring the pressure and temperature. In the two liquid injection lines the working fluid flow rate is measured. This way the mass flow rate and composition of the

working fluid can be determined in each section between points H and A. Pressure and temperature are measured separately before and after the compressor (points H^x and A^x). By changing the distribution of the injected liquid and by by-passing the separation, different variants of the compression process can be provided and measured.

On the marked measuring points temperature can be measured at 15, pressure at 6, pressure difference at 2, mass flow rate at 5 places. Beyond this the electric power demand of the compressor and the liquid pump also can be measured. The measured data can be recorded. The data-gathering device operates with checking time system that can be set in 8 steps within 0.5 - 20 minutes. Processing of the measured values is carried out in a way that an average value of 10 readings is considered one measurement. Evaluation of the data recorded on a cassette is accomplished by a computer located off-site.

Data processing takes place with two kinds of measurement evaluating programs. One of them serves for the general evaluation of the heat pump, while the other evaluates the cycle of the heat pump and the processes in the heat exchangers. While compiling the latter program the experience gained with former tests on heat exchangers were taken into consideration [16].

8. CONCLUSIONS

Based on the first experiences gained during operation the following conclusion can be drawn:

1. In the heat exchangers of the hybrid heat pump the expected gliding temperature profile develops, the measured parameters come close to the planned data.

2. The construction of the heat exchangers meets the requirements, ensuring the heat exchange in counterflow between the media, which is absolutely necessary for realizing the expected thermodynamical advantages.

3. The application of wet compression makes the operation of the screw pair of the compressor smoother and more silent than with dry compression. For commercial equipment, however, reliable accessories, first of all an improved sealing system is required.

4. The hybrid heat pump pilot plant proves its expected thermodynamical advantages concerning external media with significantly gliding temperatures.

5. The operation of the equipment is not considerably more intricate than that of a conventional heat pump and with suitable components (e.g. compressor) it may be rendered completely automatic.

REFERENCES

1 Osenbrück, A.: Verfahren zur Kälteerzeugung bei Absorptionsmaschinen. Deutsches Patent 84084 (1895).

2 Altenkirch, E.: Reversible Absorptionsmaschinen. Zeitschrift für die gesamte Kälte-Industrie, 1913. Heft 6. S. 114-119.

3 Altenkirch, E.: Die Kompressionskältemaschine mit Lösungskreislauf. Kältetechnik, 10-12/1950.

4 Rosenfeld, L. M. - Tkachov, A. G.: Refrigerating machines and apparatuses, chapter XIII., p. 448-473 (Russian). Gostorgizdat, Moscow, 1955.

5 Bošnjaković, F.: Technische Thermodynamik, II. Teil. Verlag Theodor Steinkopff, Dresden, 3. Auflage, 1960.

6 Bošnjaković, F.:Technische Thermodynamik, II.Teil. Diagramm-Mappe der Zweistoff-Gemische, Tafel IV-V. Verlag Theodor Steinkopff, Dresden, 3. Auflage, 1965.

7 Najork, H.: Kompressionskältemaschine mit Lösungskreislauf. DDR-Patent 136 176 (Anmeldung: 1976).

8 Hivessy, G. - Pecz, P.: Hybrides Kompressions-Absorptions verfahren für das Betreiben von Wärmepumpen oder Kältemaschinen. Europäisches Patent 0 021 205 (Priorität: 1979).

9 Trepp,Ch. - Ziegler,B.: Zustandsgleichung für Ammoniak-Wasser-Gemische. Klima-Kälte-Heizung, 6/1983.

10 Mučić, V. - Scheuermann, B.: Zweistoff-Kompressions-Wärmepumpe mit Lösungskreislauf (Prinzip, Konstruktion und Versuchsergebnisse). Fernwärme International Nr. 13. 1984, Heft 2., S. 79-81.

11 Bergmann, Gy. - Csaba, G. - Hivessy, G.: A heat exchange pipe for heat transfer,European patent 0 242 838 (Priority: 1986).

12 Bergmann, Gy. - Hivessy, G.: Advanced hybrid heat pump. Hungarian patent, 198 329 (Priority: 1986).

13 Bergmann, Gy. - Hivessy, G.: Hybrid heat pump - a way of increasing the heat pumping efficiency. UNO-ECE Symposium on the Long-Term Impact of Energy Efficiency Improvements, Budapest, April 1987.

14 Schäfer, W. - Funcke, H.: Kompressionswärmepumpen mit Lösungskreislauf zur Nutzung von Geothermie und Sekundärenergie für die Wärmeversorgung. Vortrag auf der Fachtagung Kältetechnik, Dresden, 1987.

15 Bergmann, Gy. - Hivessy, G.: Hybrid heat pump for hot water supply. XVIIth International Congress of Refrigeration, Vol. E, p. 250-255. Vienna, August 1987.

16 Bergmann, Gy. - Hivessy, G.: Laboratory test of heat exchangers for hybrid heat pumps. XVIIth International Congress of Refrigeration, Vol. E, p. 522-527. Vienna, August 1987.

17 Malewski, W. F.: Integrated absorption and compression heat pump cycle using mixed working fluid ammonia and water. Proceedings of the Institute for Refrigeration, 1987-88/4.

18 Bergmann, Gy. - Hivessy G.: Development of hybrid heat pumps. 2nd International Workshop on Research Activities on Advanced Heat Pumps. University of Technology Graz, September 1988.

19 Bergmann, Gy. - Hivessy, G.: The multi-stage hybrid heat pump. 2nd International Workshop on Research Activities on Advanced Heat Pumps.University of Technology Graz, September 1988.

20 Bergmann, Gy.: Fundamental principles of hybrid heat pumps (Hungarian). Hütöipar, January-March 1989, p. 5-14.

21 Hivessy, G. - Homola, T.: Energetische Bewertung der hybriden Wärmepumpe bei Wärmenutzung in Kühlhäusern. VII. Kältemaschinen--Fachseminar, Tata (Ungarn), September 1989.

22 Bergmann, Gy. - Hivessy, G.: Hybrid heat pump pilot plant. IEA Heat Pump Center Workshop on High Temperature Heat Pumps. Hannover, November 1989.

23 Ziegler, F. - Alefeld, G.: Compression-absorption heat pumps for high-temperature applications. IEA Heat Pump Center Workshop on High Temperature Heat Pumps. Hannover, November 1989.

24 Bergmann, Gy.: Hybrid heat pump (Hungarian). Energiagazdálkodás, January 1990, p. 1-8.

Chapter 4

A 'SELF-REGULATING' HEAT PUMP FOR RENEWABLE ENERGY APPLICATIONS

C L Pritchard and R E Low (University of Edinburgh, UK)

SUMMARY

Several European countries have embarked on major programmes for the development of wind power. These have been primarily directed at supplying electrical power to the grid. However, since wind power cannot be relied upon to meet base loads, the economics have been based on the resulting fuel savings at base-load stations. These calculations, not surprisingly, show wind energy in a very unfavourable light. Nevertheless, many processes may be operated with a time-variable energy input: greenhouse heating, drying and desalination are examples. Such thermal processes are particularly suited to windpower, since the wind can supply not only shaft work, but also low-grade thermal energy, particularly in the form of the latent heat of its moisture content. Thus a heat pump system that is capable of upgrading ambient thermal energy over a wide range of shaft power input could be driven by a wind turbine generator (or wind-driven compressor) to supply heat for moderate temperature thermal processes.

The use of a working fluid of variable composition in a compression heat pump makes possible the abstraction of heat at useful temperatures over a range of compressor work inputs. Such a system has been designed and constructed to use a zeotropic mixture of R113/R114, whose composition is automatically adjusted in response to variations in the available shaft power. Prelimin ary operating experience and experimnetal results from this prototype are described. For the system to be used in conjunction with windpower devices, the fluid composition must respond rapidly. A method of achieving this is described, with the aim of matching the self-regulating heat pump to wind energy applications.

INTRODUCTION

In recent years, several European countries have embarked on major programmes for the development of renewable energy sources: solar, wind- and wave-power. Although many notable successes were scored in the design, construction and operation of working prototypes, most of the impetus of these programmes was lost as oil prices tumbled in the eighties. However, programmes for the development of windpower did retain a toehold, so that at the end of the decade several countries, including Denmark, the Netherlands and the U.K., have a number of large (>250kW) wind turbines in operation, and viable programmes of wind energy development. With the dawning realisation that the life-cycle costs of nuclear-generated electricity, at around 7p/kWh, exceed the probable costs of generation from wind or wave sources; and with mounting international concern about atmospheric heating and the Greenhouse effect, programmes for the development of renewable energy sources are expected to accelerate in the 1990s.

A major factor impeding the immediate introduction of windfarms on the coasts of Northern Europe is the variable nature of the resource being tapped: the available mechanical energy in a 15 m/s wind is ten times that of a 7 m/s wind; 27 times that in a 5 m/s wind in which a large machine would begin to deliver power. This extreme variability, coupled with a rapid

Heat Pumps – Proceedings of the 4th International Conference, pp.41-48

rate of change, restricts the contribution of wind-generated electricity to the "top slice" of demand, with a zero (or small notional) contribution to base loads. Wind-generated electricity has thus been valued only at fuel costs; and in these terms is, not surprisingly, uncompetitive with base-load generation systems.

It has been suggested that wind energy would be most valuable when used directly in interruptible processes: one proposal is for the direct use of wind energy in geenhouse heating, for which the heat demand increases with the windspeed! There are however many other high-energy processes which could be powered by a variable supply: drying processes and desalination are two examples. Their comon feature is that they require a source of heat at relatively low temperature: in the range 80-120°C. Such heat may be provided directly by using the shaft work of a wind turbine for Joule or Ohmic heating; or it may be provided at much higher efficiency by utilising the (variable) shaft work input to drive a heat pump which abstracts heat from the airstream. Such a heat pump could deliver heat equivalent to up to four times the shaft work input. Alternatively, a given heat output could be provided by a heat pump-coupled wind turbine having half the diameter of one driving Joule or Ohmic heating. The enthalpy content of the airstream passing a wind turbine (or indeed of the sea passing a wavepower device) exceeds by a thousandfold the mechanical work abstractable from the stream. This makes heat pumping a very attractive adjunct to such renewable energy sources.

Heat pump development work over the last few decades has concentrated on the achievement of improvements in c.o.p., and on capacity control, in systems with a near-constant load. In conventional vapour compression heat pumps, the compressive work is done at constant rate, leading to devices which have little flexibility to respond to changes in load and/or the availability of power. It has therefore been necessary to develop a system which can produce a useful heat output at near-constant temperature over a wide range of variation in shaft work input. Such a system would make possible the effective use of wind or wavepower for a wide range of thermal processes: desalination, steam raising, drying, thermochemical processes, space heating. An economic evaluation of such a heat pump installation has indicated that economies of scale are achievable in installations rated up to 1 MW (thermal).

TECHNICAL REQUIREMENTS

The basic requirement of such a system is that it should deliver heat at a useful temperature while the power input varies over a wide range. Conventional capacity control systems have a cut-off point at around 50 percent of maximum rated power input, because of problems of compressor lubrication and declining c.o.p. at low compressor speeds. However, the effective utilisation of wind and wave power requires that useful energy be captured for a high proportion of the operating time. Studies of the power availability from these sources suggest that a turndown ratio of 10:1 is desirable if the system is not to be either 'stormbound' (overpowered) or 'becalmed' (inoperative) for much of the time. The heat source will be the ambient energy stream - wind or sea - at an effectively constant temperature. Our pilot plant design is based on upgrading ambient heat at 20°C to useful process heat at up to 100°C, for which the Carnot coefficient of performance $(cop)_C$ = 4.66.

CONCEPTUAL DESIGN OF PROTOTYPE HEAT PUMP

The design incorporates two major innovations: use of a variable-speed rotary compressor with approximately constant compression ratio of 5.5, to give a widely varying throughput; and the use of a binary zeotropic working fluid mixture whose composition may be varied continuously in response to changes in available power input. Taken together, these innovations allow the condenser outlet temperature to be maintained at a near-constant value whilst the compressor power input varies over an order of magnitude. Typically, such a power variation would be obtained over a wind speed range of 7 to 15 ms^{-1} or a wave height range of 0.5-1.5 m.

The potential advantages of using a nonazeotropic mixture as working fluid have been widely studied (Berntsson and Schnitzer 1984; Ofner et al. 1986; Patwardhan et al. 1987). Redermacher (1986) has proposed heat pump schemes utilising a solution circuit to achieve high outlet temperatures from a vapour compression system whilst retaining a low pressure ratio. Stokar and Trepp (1987) have demonstrated that such a system can give a substantial improvement in c.o.p. over a conventional single-fluid heat pump. The advantages that accrue have been shown by Patwardhan (1987) to be very sensitive to the working fluid composition . In our plant, this behaviour is exploited by adjusting the composition of the mixture in response to changes in power availability, in order to maintain an optimum c.o.p. over a range of operating conditions.

The working fluid is separated by using a two-stage condensation system with intermediate pressure reduction. The first, high pressure condenser gives a liquid rich in the heavy component; whilst the product from the lower-pressure condenser is rich in the lighter component. The use of two (or more) condensers in series enables the refrigerant mixture to be separated continuously on-line. The use of intermediate pressure reduction also permits a close temperature approach to be maintained between the condensing refrigerant vapour and the load, with the corresponding advantages of the close approach to the Lorentz cycle. In the prototype, the load is a hot water stream heated from 65 to 85°C, and temperature approaches of 5-7 degrees are maintained in the two condensers, which are welded plate units. The flowrate of the hot water system is controlled to give a constant outlet temperature of 85°C.

The choice of working fluid pair has been constrained by many aspects of the design:

i) Source temperature 15-20°C ; delivery temperature 85-100°C

ii) Relative volatility, as determined by the separation of dew and bubble points of an equimolar mixture, must be adequate to achieve good separation of the components in the two condensers in series.

iii) Stability and non-toxicity.

iv) Flexibility to allow use of CFC replacements as these become available in the requisite quantities.

v) Relatively low pressure ratio across the range of working compositions.

vi) The mixture rich in the less volatile component should have vapour pressure > 1 bara at the lowest working temperature of the system. (This is a practical measure to minimise the possibility of air ingress.)

The suitability of a large number of working fluid pairs, including CFC replacements, was determined against these criteria; the results are in Table 1.

Table 1 : Key Properties of Refrigerant Mixtures

Equimolar mixtures, assumed ideal.
Evaporator dewpoint = 20°C ; Condenser dewpoint = 85°C

Mixture	P(evap) (bara)	P(cond) (bara)	$\frac{Pc}{Pe}$	x(evap) mole fraction	x(cond)
R12/R11	1.56	9.89	6.34	0.135	0.232
R12/R114	2.73	14.80	5.42	0.256	0.353
R114/R11	1.17	7.36	6.28	0.340	0.390
R114/R113	0.61	4.70	7.72	0.168	0.255
RC318/R11	1.28	8.16	6.39	0.279	0.338
RC318/R113	0.63	5.03	7.93	0.132	0.212
R22/R11	1.63	10.75	6.57	0.093	0.172
R22/RC318	3.99	21.88	5.48	0.239	0.363
R22/R114	3.03	17.14	5.66	0.180	0.287
RC318/R123	1.16	7.94	6.78	0.239	0.324
R12/R123	1.36	9.49	6.97	0.110	0.220
R114/R123	1.06	7.12	6.72	0.301	0.376
R22/R123	1.42	10.30	7.26	0.071	0.161

The methods used for determination of the properties of mixtures are detailed by Low (1988). The most suitable mixture for this purpose is RC318/R123a; and our proposed operating cycle for this pair is shown on the P-H-z diagram, Fig.2. (However, since these refrigerants are not yet commercially available, the prototype is being run initially with a R114/R113 mixture.)

The prototype heat pump, rated at 0.3-3.3 kW power input, utilises a Rotocold R3 compressor and electrical variable speed drive, which is controlled by computer to simulate variations in power input from a renewable energy source. A further innovation is the use of high-intensity, extended surface heat transfer tubing in the evaporator, which again serves to minimise the temperature difference between source (at around 20°C) and evaporating refrigerant to between 1 and 10 degrees, depending on load.

Working fluid is stored in two vessels of 6 litres (~10 minutes') capacity, each receiving fluid from one of the condensers. A twin-headed metering pump automatically blends the rich and lean mixtures, in response to the available compressor power input: at high power, a high proportion of the less volatile component is used, to give a large temperature difference driving force for heat transfer at the condensers. At lower power inputs the condensers are underloaded, and the condenser outlet temperature can be maintained with a higher proportion of the more volatile component of the working fluid.

This system as originally designed has a rather slow response to changes in power input. This is principally because the storage vessels are well-mixed, but also partly a result of the liquid holdup in the evaporator, and partly because the degree of separation achieved by the two-condenser system is itself a function of throughput. The response time of ~10 minutes is, however, adequate for wave-driven systems, in which the energy intensity of incoming waves changes on a timescale of hours rather than minutes. The proposed adaptations required to give a rapid response in windpowered systems are considered later.

OPERATION OF PROTOTYPE

Figure 1 is a detailed flow diagram of the prototype. Refrigerants stored in pressure vessels (T4,T3) are blended by metering pump (P1,P2) and delivered to the evaporator (E1) which is heated by a water circulation system at ~20°C. (This source fluid also collects heat from the vessel coolers H1,H2.) The vapour from the evaporator may be mixed with flash vapour from either condensate stream and is superheated in a double-pipe exchanger X1. As well as contributing to the thermal economy of the system, this recirculation of flash vapour may be used to improve the separation of refrigerants achieved in the storage vessels. The mixed vapour enters the compressor (J1) whose efficiency may be improved by liquid injection. The compressor speed is under computer control, and may be used to simulate the power variations obtained when wind- or wave- power drive the compressor.

Compressed vapour is condensed in two stages (C1,C2) : the pressure in the first is controlled according to load, to give the requisite condensation temperature; whilst that in the second is controlled to complete the condensation of remaining vapour which is rich in the more volatile component. Liquid from each condenser is separately returned to storage (T4,T3) via subcoolers (H2,H1). These subcoolers are bypassed when flash vapour return is used. The prototype has provision for liquid injection into the first condenser (C1), which provides the possibility of further increases in the range of operation and in overall c.o.p.

Data collection and control are carried out by an IBM-XT compatible microcomputer, using an IEEE interface card and CIL monitors for thermocouple inputs, and a proprietary analogue interface card for the other analogue inputs and control outputs. Inputs include flows of source and load fluid; levels in the separation pots (T1,T2); pressures in evaporator, condensers and storage; and compressor power consumption. Outputs control the compressor speed, condenser pressures, condensate levels, load temperature, the blending pump rates and the flash vapour return. The composition of refrigerant in each storage vessel is estimated from simultaneous readings of storage temperature and pressure, and may be checked off-line by densitometer or chromatograph.

The control philosophy is as follows:-

- Flow control of the load fluid is used to maintain the load temperature at the desired level.
- Total flow of refrigerant is controlled by the refrigerant level in the evaporator (which varies in response to compressor speed.)
- Target composition of refrigerant is set by the desired load temperature and given compressor speed: this algorithm is stored in the computer. The actual composition is determined from temperature and pressure at compressor inlet, and is controlled by the flow ratio at the metering pump heads.
- Condenser pressures are controlled so as to divide the condensation load between the two condensers.

The complete thermodynamic cycle for this system is superimposed on the P-H-z diagram, Figure 2. It has been drawn for a working fluid comprising an equimolar mixture of RC318 and R123a, operating between pressures of 1.2 and 8.8 bara (C1) and 7.9 bara (C2). The numbered points on Fig.2 correspond to the streams of the same number on Fig.1, and the diagram shows the split in composition achievable using the two-condenser system.

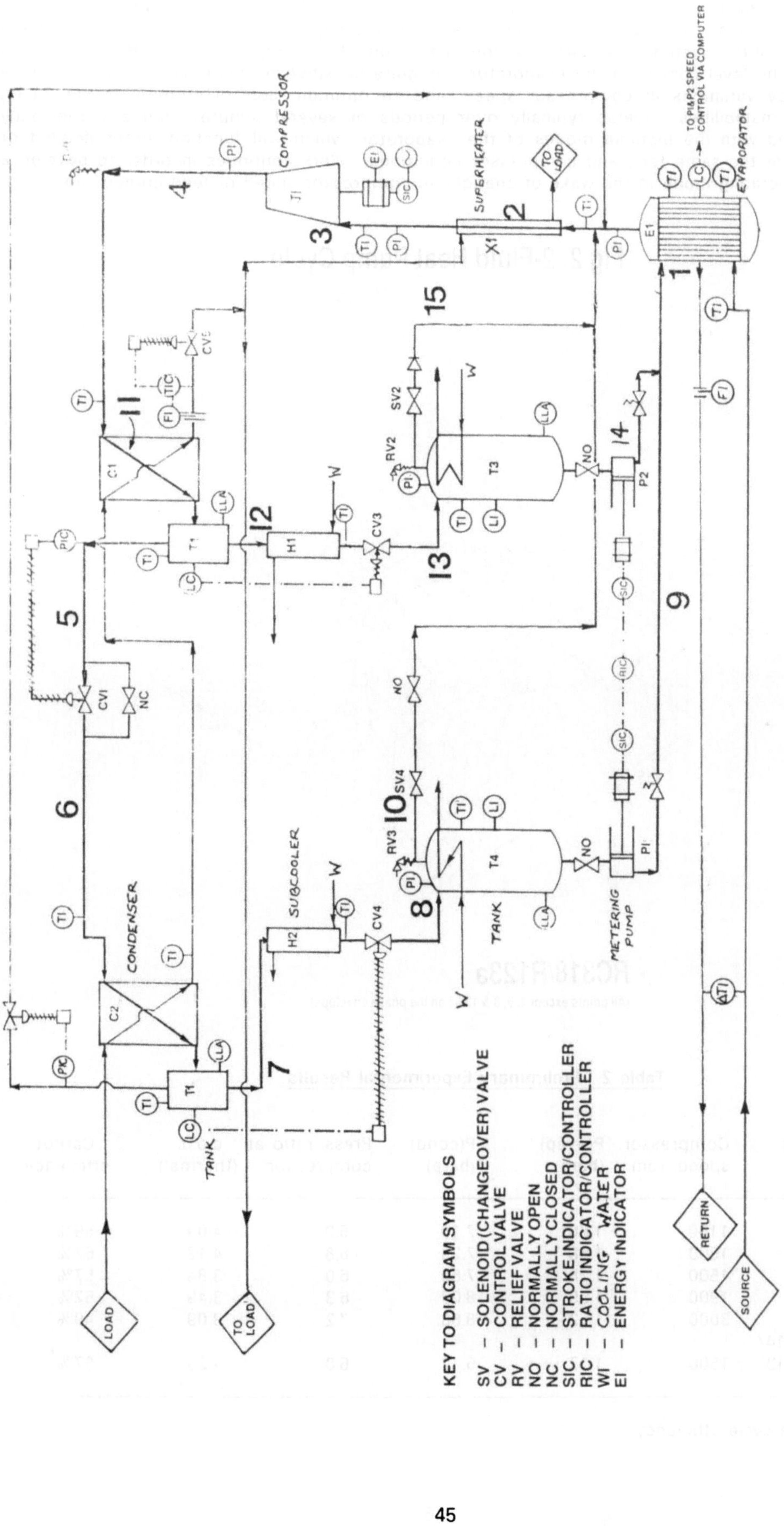

Fig. 1 Flowsheet for 2-Fluid Heat Pump

EXPERIMENTAL RESULTS

The prototype has been operated at steady throughputs with a binary mixture of R113/R114 as working fluid. Some preliminary performance data are given in Table 2. It has been found that the control system - in particular the separation of working fluid in condensers C1 and C2, and the level control in the evaporator - is quite sensitive to disturbances such as those caused by variations in compressor speed; and an optimum control strategy is still being sought. Instabilities develop typically over periods of several minutes, and are principally associated with the bistable modes of the evaporator, which will function either flooded or half-full at the same feed and compressor conditions. Work continues in order to develop a stable operating mode in the wake of changes in compressor speed or feed composition.

Fig 2: 2-Fluid Heat Pump Cycle

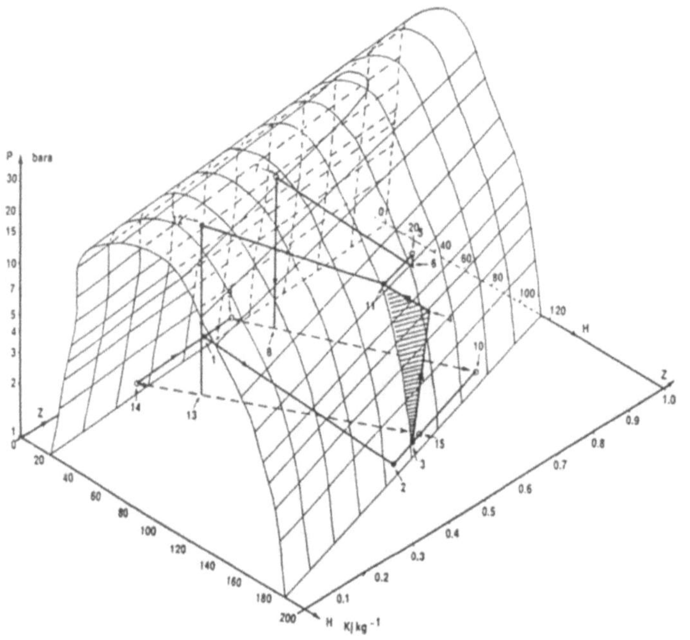

RC318/R123a

(All points except 4, 6, 8 & 13 lie on the phase envelope)

Table 2 : Preliminary Experimental Results

Working Fluid	Compressor speed, rpm.	P(evap) (bara)	P(cond) (bara)	Press. ratio at compressor	c.o.p. (thermal)	Carnot efficiency
R114	1150	1.80	7.10	5.0	4.09	59%
R114	1500	1.58	7.23	6.8	4.12	67%
R114	1500	1.78	7.68	6.0	3.84	57%
R114	1900	1.89	8.09	6.3	3.49	52%
R114	3000	1.52	8.68	7.2	3.09	49%
60% R114/ 40% R113	1500	1.17	5.12	6.0	4.25	57%*

* Lorenz cycle efficiency.

IMPROVEMENTS TO SYSTEM RESPONSE

In order to develop a system capable of responding to rapid changes in power availability, such as those developed by a low-inertia wind turbine in gusty conditions, it is necessary to minimise the working fluid holdup in condensate storage and subcoolers; and it is desirable to simplify the blending system itself to minimise liquid holdup. We propose to achieve this by storing the "rich" and "lean" working fluids at their respective condensing pressures, with a "last in, first out" storage system to permit rapid changes in composition in the circulating fluid. Fig. 3 is a simplified diagram of the system showing this storage and blending system. Liquid from the appropriate storage vessel (T1 or T2) is selected by solenoid valve S1 and admitted to the evaporator under level control. Vapour flashing off at the control valve is routed by S2 to the appropriate part of the evaporator. When the more volatile component is admitted, S2 delivers the 2-phase mixture to the vapour space of the evaporator; hence the flashed vapour is fed directly to the compressor inlet. About 38 percent of lights are flashed, and this contributes significantly to a fast response time when power input is decreasing. When the less volatile mixture is admitted, S2 directs it to the base of the evaporator, to reduce the separation of lights from this stream. The holdup in the evaporator itself has been minimised by the use of highly extended heat transfer surface, giving a very large surface-to-volume ratio. The modifications described would enable the prototype to operate with an "active" working fluid holdup of around 0.9 kg, corresponding to 14 seconds of compressor throughput at full rate. Further reductions in the response time could be achieved by direct liquid injection into the compressor and/or condenser.

SCALE-UP

The prototype has been designed around the smallest available variable speed rotary compressor. With its constant pressure ratio, this exhibits a roughly linear power-speed relationship. This is in contrast to the proposed energy sources for the system: the power available from the wind is proportional to the **cube** of the windspeed; whilst that available from deep water waves is inversely proportional to wave frequency. A full-scale device would utilise a hydraulic power transmission system between the prime mover and the compressor to enable the characteristics of the incoming energy stream to be matched to those of the compressor. The conversion of wind and wave energy to shaft work is well documented, and is not considered further here.

CONCLUSION

A prototype "self-regulating" heat pump has been built and operated over a range of power inputs. The technology offers promise as a means of utilising wind/wave energy to supply heat at useful temperatures for drying, space heating or desalination.

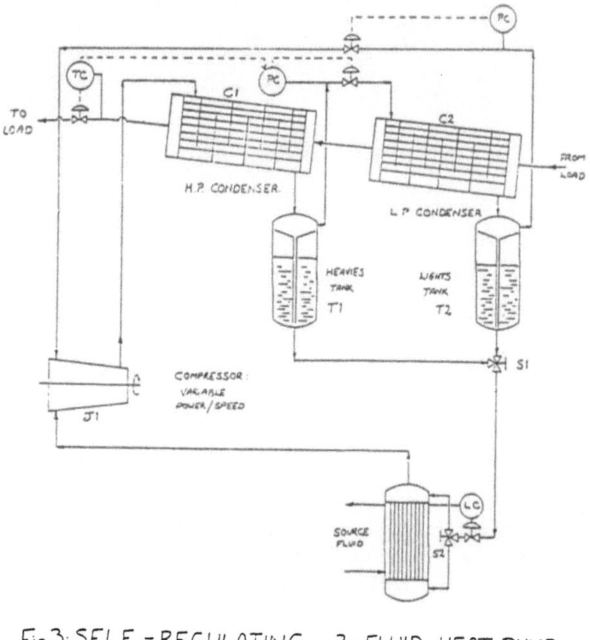

Fig 3: SELF-REGULATING 2-FLUID HEAT PUMP: SIMPLIFIED FLOW DIAGRAM

REFERENCES

Berntsson, T. & Schnitzer, H. 1984. "Some technical aspects on nonazeotropic mixtures as working fluids", Proc. 2nd Intl. Symp. on the Large Scale Applications of Heat Pumps, paper A1, B.H.R.A. (U.K.)

Low, R.E. 1988. "Techniques for thermodynamic property prediction with reference to microcomputer application", Proc. 2nd Intl. workshop on Research Activities on Advanced Heat Pumps, Graz.

Ofner, H., Schnitzer, H. & Moser, F. 1986. "The economy of nonazeotropic refrigerant mixtures in compression heat pumps", J. Heat Recovery Systems **6**, (4), 313-321, 1986.

Patwardhan, V.R., Devotta, S. & Patwardhan, V.S. 1987. "Performance Characteristics of a water-water heat pump using nonazeotropic mixtures", Proc. 3rd Intl. Symp. on the Large Scale Application of Heat Pumps, paper B4, B.H.R.A. (U.K.)

Radermacher, R. 1986. "Advanced versions of heat pumps with zeotropic refrigerant mixtures", ASHRAE Transactions **2977**,

Stokar, M. & Trepp, Ch. 1987. "Compression heat pump with solution circuit, Part 1: design and experimental results", Int. J. Refrig. **10**, 87-96.

Chapter 5

EXPERIMENTAL RESULTS OF A DOUBLE-LIFT COMPRESSION-ABSORPTION HEAT PUMP

F Ziegler and G Hämmer (Technical University, Munich, Germany)

Synopsis:

A compression heat pump with an integrated solution circuit which increases the temperature lift is presented. A breadboard-scale unit operating with the working pair R22/DMETEG has been tested successfully. The heat pump is capable of heating tap water to more than 100°C and simultaneously providing refrigeration at -20°C. The operational characteristics and the performance of the cycle will be reported.

Nomenclature:

A	Absorber
C	Condenser
COM	Compressor
E	Evaporator
EV	Expansion Valve
G	Generator
SHX	Solution Heat Exchanger
SP	Solution Pump
↓ ↑ ←	Input or Output of Heat or Power
Q_r	Refrigeration Capacity
Q_h^l	Low temperature heat output
Q_h^h	High temperature heat output
W	Shaft work of compressor including solution pump
T_h	Temperature at the outlet of the heat sink
ΔT	internal temperature lift of the cycle

INTRODUCTION

In this paper first experiments with a double-lift compression-absorption cycle will be reported. With this two-stage cycle a temperature lift of more than 100 K has been reached although only one conventional compressor has been used. The cycle and the experimental

Heat Pumps – Proceedings of the 4th International Conference, pp.49-58

setup are described. The cycle can be operated easily in a stable and reliable way. Experiments have been performed producing refrigeration at -10°C and -20°C with simultaneous heat output in the temperature range of 50°C to above 100°C. Hence the compression-absorption heat pump offers the possibility of producing an output of heat at a high temperatur even if refrigeration at quite a low temperature is required. By this means a significant amount of primary energy can be saved.

A heat pump cycle is called a compression-absorption cycle when a solution circuit as known from absorption devices is used in a compression heat pump or when a compressor or fan is used in an absorption heat pump or heat transformer system. The most simple compression-absorption cycle was presented in 1895 by Osenbrück. It has been analysed by Altenkirch (1950) in detail. The cycle is shown schematically in Fig. 1 together with a conventional cycle in a pressure-temperature plot. Condenser C, expansion valve EV and evaporator E of the conventional compression cycle are replaced by a solution circuit consisting of an absorber A, a solution heat exchanger SHX, an expansion valve EV, a generator G and a solution pump SP. The condensation process is replaced by absorption and the evaporation is replaced by regeneration.

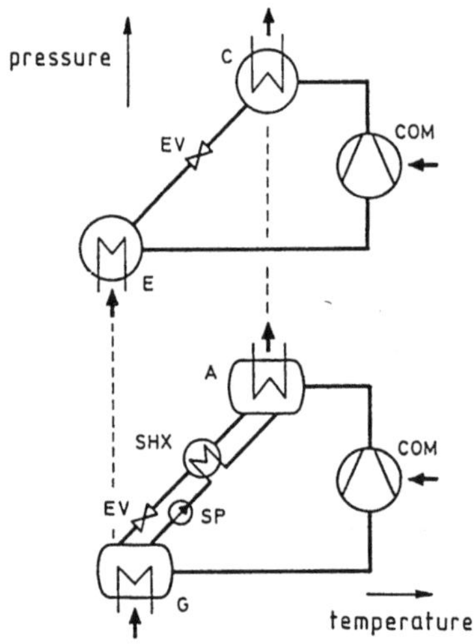

Fig. 1: Conventional compression heat pump and single-stage compression-absorption cycle in a pressure-temperature plot (Abbreviations see Nomenclature)

There are a lot of other possibilities to combine elements of compression and absorption cycles leading to machines with interesting new features (Alefeld 1982). Yet only few of these cycles have been analysed in detail; the state of art is reviewed by Morawetz (1989) and Ziegler and Alefeld (1989). Several highly efficient cycles have been analysed by Alefeld and Ziegler (1985), Ahlby (1987), Radermacher and Howe (1988), Ziegler (1988) and Bergmann and Hivessy (1988). Here we will report on a cycle with a high temperature lift (Alefeld 1982, Pourreza-Djourshari and Radermacher 1986, Radermacher 1987, Ahlby 1987, Rane, Radermacher and Herold 1989).

THE DOUBLE-LIFT CYCLE

The double-lift cycle is shown in a pressure-temperature plot in Fig. 2. The condenser C of a conventional compression cycle E-COM-C rejects its heat to the generator G of a single-stage compression-absorption cycle G-COM-A; however both cycles are integrated and thus operate with the same compressor. So with single-stage compression heat can be pumped from the evaporator E up to the absorber A, which is in the order of twice the temperature lift attainable with a conventional compression cycle. This is the reason for calling it "double-lift" cycle.

Because the cycle consists of two stages we will describe them one after the other although the two stages are integrated. Temperatures and pressures are typical data for the experimantal plant described below.

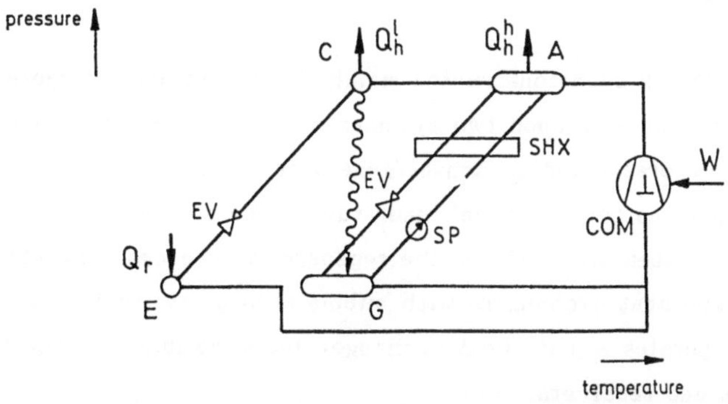

Fig. 2: Flow scheme of the double-lift compression-absorption heat pump in a pressure-temperature plot (Abbreviations see Nomenclature)

The first stage is the conventional one. In the evaporator E the refrigerant condensate is vaporized producing refrigeration at -20°C. The refrigerant vapour is sucked off by the compressor COM at 3 bar and discharged to the high pressure line at e.g. 20 bar. Part of the refrigerant vapour is absorbed in the absorber A which belongs to the second stage. The remaining fraction of the vapour is separated from the absorbing solution and flows on to the condenser C, where it is liquefied at 50°C. The condensate is being expanded in a thermostatic expansion valve EV into the evaporator.

The condenser can be cooled by cooling water as usual in conventional machinery. To attain the double lift, however, it has to be cooled by the generator. So within the condenser C heat rejected from the first stage is being transferred to the second stage. In the generator G it regenerates the so called strong solution at e.g. 45°C. The pressure of the generator G is the same as the pressure of the evaporator E. So after being separated from the weak solution the vapour produced during regeneration is mixed with the vapour coming from the evaporator E. The vapour then is pressurized by one common compressor COM. The regenerated weak solution is pumped by the solution pump SP through a solution heat exchanger SHX, where it is heated, to the absorber A, which is operated at discharge pressure. There it absorbes at e.g. 100°C part of the vapour leaving the compressor. The heat of absorption is rejected to the heat sink. The strong solution is separated from the remaining vapour which flows on to the condenser. The solution is cooled in the solution heat exchanger SHX and is expanded into the generator to be regenerated.

So heat can be pumped in two steps, first from the evaporator to the condenser and then from the generator to the absorber. A high temperature lift is achieved with only one compressor at a moderate pressure ratio.

A breadboard unit of this cycle working with NH_3 as refrigerant and NH_3/H_2O as absorbing solution has been built by Rane, Radermacher and Herold (1989). The machine is able to be operated in more complicated modes as well.

A unit of the same size of about 5 kW cooling capacity has been built and operated at the Technische Universität München. R22 is used as refrigerant and R22/DMETEG (Yaron, Borde 1984, Ehmke 1985) as absorbing solution. We will report about the first experiments with this double-lift compression-absorption heat pump.

THE EXPERIMENTAL SETUP

The experimental plant has been constructed mainly from standard refrigeration equipment in copper. The compressor is an open two cylinder piston compressor with water-cooled cylinder head. The evaporator and condenser/generator are coaxial tube heat exchangers. The condenser incorporates three parallel inner tubes. One of them is used for external cooling, the other two tubes are used for the regeneration process. The absorber consists of two helical coil type heat exchangers with bubble absorption inside the helical coil. The solution circuit incorporates a plate heat exchanger and a membrane pump. Not shown in Fig. 2 are separators and receivers.

The experimental setup is equipped with temperature probes, pressure gauges and flow meters in the cycle itself as well as in the heat sink and heat source circuits. Additionally the consumption of electrical power of the drive of compressor and solution pump are measured.

The heat source for our compression-absorption heat pump is a circuit with a methanol-water mixture. It is heated electrically.

In the experimental setup it is possible to cool the condenser externally in addition to the internal heat exchange between the condenser and the generator. Thus the amount of heat Q_h^l transferred from the condenser to a first heat sink, in relation to the heat Q_h^h rejected from the absorber to a second heat sink at a higher temperature, can be varied from 0 to 1. By this means for a given refrigeration duty Q_r the amount of high temperature heat Q_h^h is adjustable with an optimal performance. The two circuits for the heat sink can be operated in series as well.

RESULTS

The first experimental results for the performance of the compression-absorption heat pump are shown in Fig. 3 to 6. The efficiency of the driving motor of the compressor has been measured in preruns and is taken into account as well as the loss of the belt drive. For the prime mover of the solution pump a relatively high efficiency of 70% has been assumed.

This approximation is good enough, because the power consumption of the pump is less than 10% of the compressor work. Hence an error in calculating the pump work does not change our results significantly. Because we are taking into account the losses of the prime movers with these corrections, the COPs shown are related to the shaft work of both compressor and solution pump.

Experiments with variable high temperature heat output are shown in Fig. 3. The COPs for refrigeration Q_r/W (+), high temperature heating Q_h^h/W ( ) and low temperature heating or reject heat Q_h^l/W (Δ) are plotted against the ratio of high temperature heating and total heat output $Q_h^h/(Q_h^h + Q_h^l)$. The heat pump cools the methanol-water mixture from about -15°C to -20°C. The first heat sink with the load Q_h^l is cooling water with 13°C inlet and about 35°C outlet temperature. The second heat sink with the load Q_h^h is a water circuit with about 60°C inlet and 80°C outlet temperature.

When the ratio $Q_h^h/(Q_h^h + Q_h^l)$ equals 0 the machine works as a conventional compression refrigerator with a COP Q_r/W of about 1.7. With an increasing amount of heat Q_h^h pumped to the high temperature heat sink the COP for high temperature heating Q_h^h/W rises from 0 to its maximum value of about 1.2 at $Q_h^h/(Q_h^h + Q_h^l) = 1$. At the same time the reject heat Q_h^l/W falls to 0. The COP for refrigeration Q_r/W drops according to the laws of thermodynamics. It reaches 0.6 when the ratio $Q_h^h/(Q_h^h + Q_h^l)$ equals 1. Then the total temperature lift attains about 100 K.

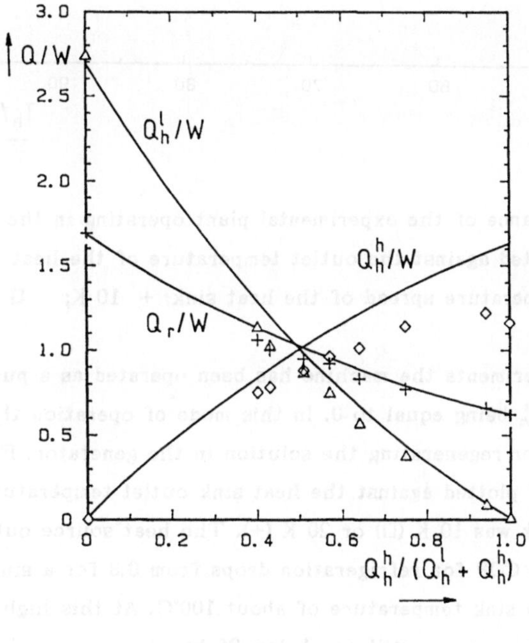

Fig. 3: Performance of the experimental plant operating with two heat sinks
+ Q_r/W; Δ Q_h^l/W; Q_h^h/W

The curves in Fig. 3 have been derived using a very simple model for calculation of the cycle. The values of the coefficient of performance for refrigeration Q_r/W in the mode with $Q_h^h/(Q_h^h+Q_h^l)$ equal 0 and 1 respectively have been used as fit parameters. The curves are in fairly good agreement with the measured data as far as the ratios Q_r/W for refrigeration and Q_h^t/W for heat output at low temperature are concerned. The experimental results for the high temperature heat output Q_h^h/W are lower than calculated due to the insulation loss of the device at high temperatures.

It has to be pointed out that the performance of the compression-absorption cycle is limited by the COP of 1.7, which is reached by the conventional part of the machine. Due to a better design of the compressor or larger heat exchangers this COP would be improved. Then all coefficients of performance of the compression-absorption cycle would increase as well.

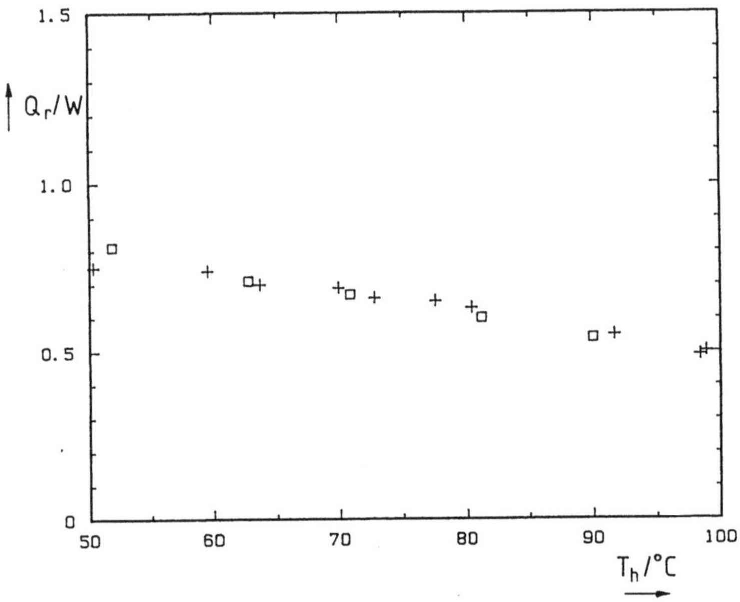

Fig. 4: Performance of the experimental plant operating in the double-lift mode, plotted against the outlet temperature of the heat sink.
Temperature spread of the heat sink: + 10 K; □ 20 K

In a second set of experiments the machine has been operated as a pure double-lift cycle with the reject heat Q_h^t being equal to 0. In this mode of operation the total heat of condensation is used for regenerating the solution in the generator. Figure 4 shows the COP for refrigeration Q_r/W plotted against the heat sink outlet temperature. The temperature spread of the heat sink was 10 K (□) or 20 K (+). The heat source outlet temperature was -20°C as in Fig. 3. The COP for refrigeration drops from 0.8 for a sink temperature of about 50°C to 0.5 for a sink temperature of about 100°C. At this high gross temperature lift of 120 K the condenser pressure still was below 25 bar.

The relatively low coefficient of performance in this mode of operation is caused by the high temperature lift which is attained. In Fig. 5 the same data as in Fig. 4 are shown in order to clarify this statement. The COP for refrigeration Q_r/W is plotted against the difference between the absorber temperature and the evaporator temperature. This temperature difference is the internal gross temperature lift of the machine. It is in the range of 100 to 130 K. Additionally some data of the pure compression cycle are shown; with a temperature difference between condenser and evaporator of 50 to 60 K the

coefficient of performance is between 1.7 and 2.4. The curve shows the performance as predicted from the data of the manufacturer of the compressor. This plot shows that the performance of the compression-absorption heat pump is quite good, if it is related to the high temperature lift which is attained.

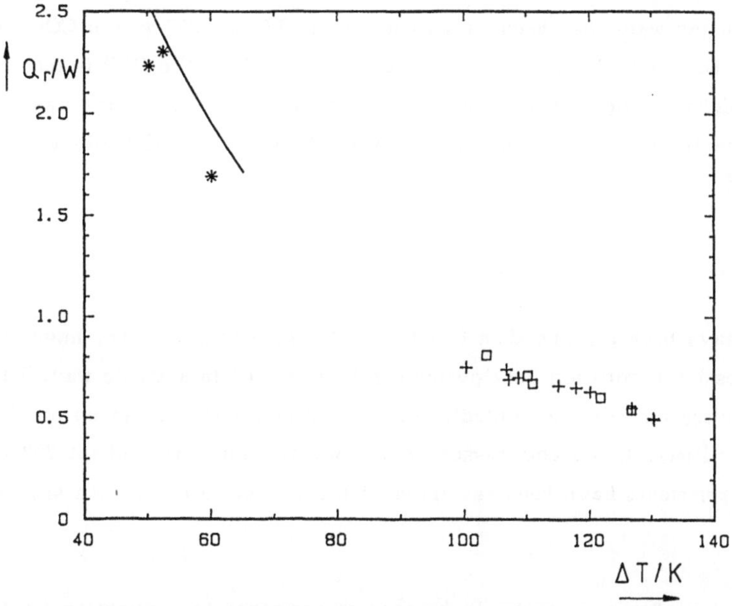

Fig. 5: Performance of the experimental plant operating in the double-lift mode, plotted against the internal gross temperature lift
Temperature spread of the heat sink: + 10 K; □ 20 K
Conventional compression cycle: ✱

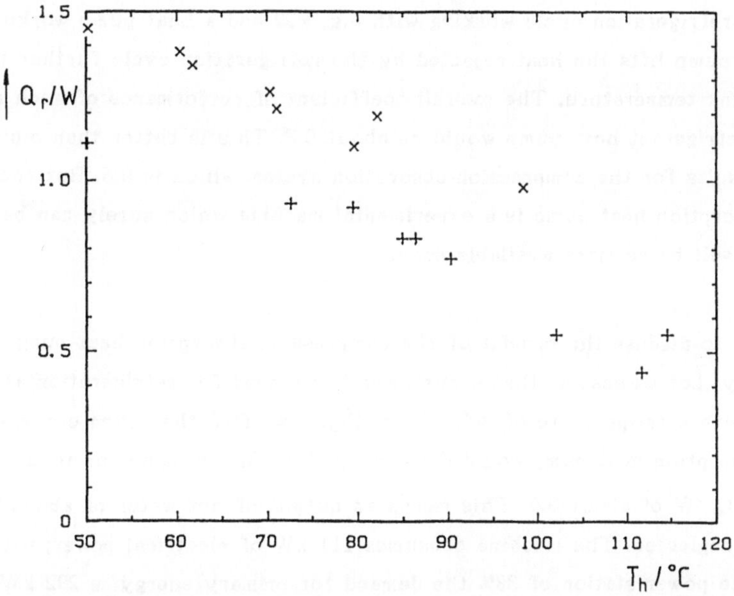

Fig. 6: Performance of the experimental plant operating with two heat sinks in series, plotted against the outlet temperature of the heat sink
Outlet temperature of the heat source: + -20°C; × -10°C

In the third set of experiments the two heat sinks have been coupled in a way, that tap water with 13°C is warmed up in three stages. It first cools the condenser, then the compressor head and finally the absorber. In Fig. 6 the COP for refrigeration Q_r/W is plotted against the heat sink outlet temperature. The heat source outlet temperature was -10°C (×) or -20°C (+). In this mode of operation at a heat source outlet temperature of -20°C, pressurized tap water has been warmed up from 13°C to 80°C with a COP for refrigeration of about 0.9. If refrigeration demand is only -10°C the COP is increased to about 1.1. The highest temperature of the heat sink outlet reached so far was 115°C at a heat source temperature of -20°C. This gross temperature lift of 135 K was attained with a COP of about 0.5.

DISCUSSION

The first tests have been performed in fall 1989. The experiments so far have shown that two-stage compression-absorption heat pumps can be operated in a stable way. The cycle is largely self-adjusting and easy to control. Startup and shutdown causes no problems. However due to oil migration a compressor breakdown happened after about 700 hours running time. Experiments have been restarted with a new compressor with slightly larger displacement.

The coefficient of performance seems to be poor as compared to conventional refrigeration machinery, but it has to be kept in mind that a large temperature lift is obtained. If a temperature lift from a heat source of -20°C to a heat sink of 80°C had to be attained with a single stage compression cycle working with e.g. R114, the COP would be about 0.5. However this cycle is technically not feasible due to the high compression ratio required. So probably a cascaded two-stage system as known from low temperature refrigeration would be used, with a refrigeration cycle working with e.g. R22 and a heat pump working with e.g. R114. The heat pump lifts the heat rejected by the refrigeration cycle further up to the required heat sink temperature. The overall coefficient of performance of such a cascaded two-stage two-refrigerant heat pump would be about 0.8. This is better than our experimental results for the compression-absorption system which is 0.6. But the presented compression-absorption heat pump is a experimental machine which surely can be improved. Moreover R114 will be no more available soon.

Finally we want to discuss the benefit of the compression-absorption heat pump for the saving of energy. Let us assume that a custumer has a need for refrigeration at -20°C and for hot water with a temperature of 80°C. From Fig. 6 we find that even our experimental compression-absorption heat pump could deliver e.g. 100 kW refrigeration with a coefficient of performance Q_r/W of about 0.9. This means an output of hot water of about 211 kW if heat losses are neglected. The machine consumes 111 kW of electrical power; with an efficiency of the power station of 38% the demand for primary energy is 292 kW.

If the refrigeration capacity would be delivered by a conventional unit with a COP of 2.25 it would consume 44 kW electrical energy and thus 117 kW primary energy. The demand of 211 kW for hot water has to be satisfied with a directly fired boiler with an efficiency of e.g. 85%. Its primary energy consumption then amounts to 248 kW. The primary energy which has to be supplied for refrigeration and production of hot water in this case is 365 kW. Hence 20% in primary energy could be saved using the experimental compression-absorption heat pump.

CONCLUSION

We have shown so far that a compression-absorption heat pump can be used for refrigeration at -20°C with simultaneously providing hot water with a temperature up to 115°C. This is achieved with a conventional single-stage piston compressor working below 25 bar. The performance of the experimental plant is at least comparable to that of conventional systems.

If a customer has a demand for refrigeration as well as for hot water with the installation of such a double-lift compression-absorption heat pump a significant amount of primary energy can be saved.

Although the possibility of combining absorption and compression technology has been known for a long time only few cycles have been analysed and applied. It was in the last decade that experiments have been restarted. The results are very encouraging. Using well-known refrigerants with low or no ozone depletion and global warming potential several heat pumps have been realized. They are able to work well above 100°C and reach temperature lifts of more than 100 K.

Compression-absorption cycles not only help in saving energy when used as high temperature heat pumps, but also will be applied for refrigeration and air-conditioning in the future.

References:

Ahlby, L, 1987, Compression/absorption cycles for large heat pumps - system simulations, Licentiate dissertation, Chalmers Tekniska Högskola, Göteborg.

Alefeld, G, 1982, Kompressions- und Expansionsmaschinen in Verbindung mit Absorberkreisläufen, Brennstoff-Wärme-Kraft, 34 (3) 142.

Alefeld, G, Ziegler, F, 1985, Advanced Heat Pump and Air-Conditioning Cycles for the Working Pair H2O/LiBr: Domestic and Commercial Applications, ASHRAE Transactions, 91 (2).

Altenkirch, E, 1950, Die Kompressionskältemaschine mit Lösungskreislauf, Kältetechnik, 2 (10) 251, (11) 279, (12) 310.

Bergmann, G, Hivessy, G, 1988, The Multi-Stage Hybrid Heat Pump, Proc. 2nd Int. Workshop on Research Activities on Advanced Heat Pumps, Graz.

Ehmke, H J, 1985, Forschungsbericht des Deutschen Kälte- und Klimatechnischen Vereins, Stoffsysteme für Absorptionswärmepumpen - Experimentelle Bestimmung thermophysikalischer Eigenschaften von Lösungen der Kältemittel Methylamin, Ammoniak und Monochlordifluormethan (R22), Stuttgart.

Morawetz, E, 1989, Sorption-Compression Heat Pumps, Int. J. of Energy Research, 13, 83.

Pourreza-Djourshari, S, Radermacher, R, 1986, Calculation of the performance of vapour compression heat pumps with solution circuits using the mixture R22-DEGDME, Int. J. of Refrigeration, 9, 245.

Radermacher, R, 1987, Vapour compression heat pump cycle with desorber/absorber heat exchange, Proc. of the 17th Int. Congr. of Refrigeration, Wien, B, 1061.

Radermacher, R, Howe, L, 1988, Internal Combustion Engine Driven Heat Pump Cycles, Paper presented at the Absorption Experts Meeting, Dallas.

Rane, M V, Radermacher, R, Herold, K E, 1989, Experimental investigation of a single stage vapor compression heat pump with solution circuit, Paper presented at the Winter Annual Meeting of the ASME, San Francisco.

Yaron, I, Borde, I, 1984, Working fluids for heat pumps, in VDI-Berichte 539, ORC-HP-Technology, VDI-Verlag, Düsseldorf.

Ziegler, F, 1988, Advanced Compression Absorption Cycles, Paper presented at the Absorption Experts Meeting, Dallas.

Ziegler, F, Alefeld, G, 1989, Compression Absorption Heat Pumps for High Temperature Applications, Proc. of the IEA Heat Pump Center Workshop "High Temperature Heat Pumps", Hannover.

Chapter 6

THE COMPRESSION-ABSORPTION CYCLE: A HIGH-TEMPERATURE APPLICATION

L Åhlby and D Hodgett (Chalmers University of Technology, Gothenburg, Sweden)

SUMMARY

A comparison of the performances of a compression/absorption cycle using the two mixtures NH_3/H_2O and NH_3/H_2O-60% LiBr shows an improvement in COP for the ternary working fluid. The improvement, however, is small; for the examples chosen it is about 5%. The advantages of the ternary mixture, such as a lower vapour pressure, a higher heat of vaporization and a reduced specific heat of solution, compared to the binary NH_3/H_2O, are to a large extent eliminated by the increase of the recirculation flow caused by the addition of salt. Another effect of the increase in solution flow is a change in the location of the optimum operating point for the ternary compared to the binary mixture.

NOMENCLATURE

A	Heat exchanger area (m^2).
COP	Coefficient of performance = $Q_{abs} / (W_{comp} + W_{pump})$ (-).
ΔT_{abs}	Temperature gradient in the absorber. Defined as the saturation temperature change during absorption = $T_3 - T_4$ (K).
F	Solution recirculation flow in the cycle (kg solution/kg vapour).
π	Pressure ratio across the cycle = P_{abs} / P_{des} (-).
P_{abs}	Absorber pressure (bar).
P_{des}	Desorber pressure (bar).
Q_{abs}	Absorber heat output (W).

Heat Pumps – Proceedings of the 4th International Conference, pp.59-68

SCD	Specific compressor displacement (m^3/MJ heat output).
T_{sink}	External sink temperature (°C).
T_{source}	External source temperature (°C).
U	Heat transfer coefficient (W/m^2 K).
W_{comp}	Compressor work (W).
W_{pump}	Pump work (W).

1. INTRODUCTION

The impact of the use of a ternary instead of a binary mixture as working fluid for a compression/absorption cycle is discussed. As an example of a ternary mixture ammonia/water-lithium bromide, NH_3/H_2O-LiBr, is considered. The performance of the cycle is calculated for this mixture and for the binary NH_3/H_2O for comparison.

Previously the performance of the compression/absorption cycle, shown in Fig. 1, has been compared for the two mixtures NH_3/H_2O and H_2O/LiBr (Åhlby 1987). From the point of view of coefficient of performance, COP (defined as the heat output from the absorber divided by the work input to the compressor and pump), the salt mixture was superior to NH_3/H_2O. On the other hand, the low vapour pressure of water results in enormous vapour volumes and therefore the size of the compressor becomes very large at the temperatures of interest.

A list of requirements on the working fluid best suited for the compression/absorption cycle is given in the same reference (Åhlby 1987). Some of these are in agreement with both NH_3/H_2O and H_2O/LiBr, but on several points these two fluids are opposites. The difference in vapour pressure for the two fluids illustrates this. A conclusion from the work cited was that NH_3/H_2O is the better working fluid and was therefore considered in later work (Åhlby, to be published). It might well be, however, that a mixture exhibiting properties in between those of the two binaries is of interest for the compression/absorption cycle. As an example of such a mixture, NH_3/H_2O-LiBr was chosen for further studies to determine whether a ternary mixture shows advantages over a binary for this particular system.

2. NH_3/H_2O-LiBr AS WORKING FLUID FOR THE COMPRESSION/ABSORPTION CYCLE

The ternary mixture ammonia/water/lithium bromide has previously been studied by, for example, Radermacher (1981). Recently additional measurements on the mixture have been presented by Zimmerman (1989). In the calculations presented below, data for the ternary mixture are taken from Radermacher (1981).

It has previously been shown by Åhlby (1987) that the optimal operating condition for the compression/absorption cycle is always found at the highest acceptable system pressure. This allows a higher proportion of volatile component in the solution and therefore minimizes the amount of solution circulated per unit vapour evaporated. The vapour pressure decrease caused by the addition of salt to NH_3/H_2O means the cycle can be operated either at lower absolute pressures at the desired temperatures or at the same pressure level, as in the case of NH_3/H_2O, but allowing a higher proportion of NH_3 in the solution. Figure 2 shows the Dühring diagrams for (a) NH_3/H_2O and (b) NH_3/H_2O-60% LiBr.

This change in operating conditions as salt is added looks, however, better for the ternary mixture than is actually the case. Both the ammonia concentration and the salt concentration are defined in aqueous solution. In these calculations, the proportion of salt is held constant at 60% by weight. Due to this definition, the Dühring diagram for the ternary mixture gives the solution field for a mixture changing from pure NH_3 in the low temperature end to H_2O/60% LiBr in the high temperature end. The higher NH_3 concentration in solution obtained for the ternary mixture at given temperature and pressure therefore, corresponds not to a smaller mass flow of solution but to a larger one. Even though a large recirculation flow is shown to be a disadvantage for the compression/absorption cycle (Åhlby, to be published), the vapour pressure decrease gives another advantage: the vapour generated in the desorber will be richer in NH_3. It is hard to say exactly what the problems would be when compressing a vapour mixture of ammonia and water. If an oil-lubricated compressor were used, there might be separation problems. In any case, it must be advantageous to generate a vapour as rich in ammonia as possible.

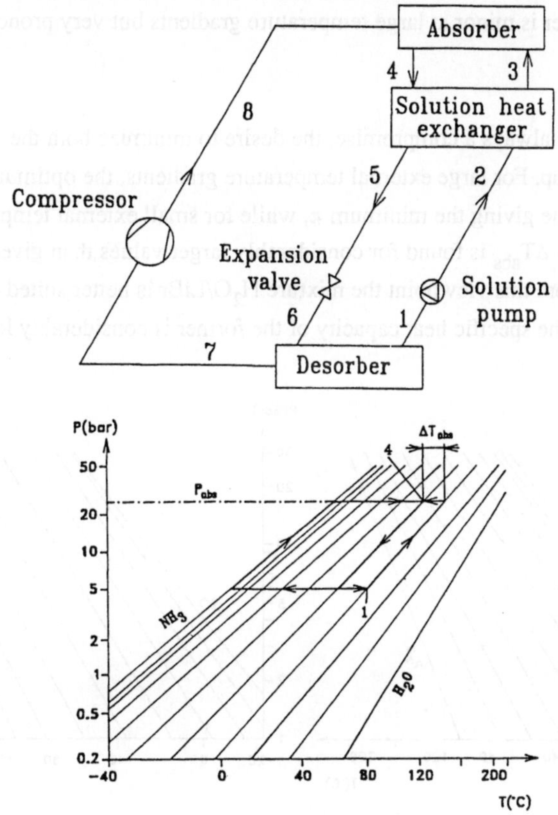

Fig. 1. Principal diagram for the compression/absorption cycle.

Figure 2 also illustrates the difference in latent heat of vaporization of the two mixtures. The steeper vapour pressure curves of NH_3/H_2O-60% LiBr correspond to a larger heat of vaporization, which is advantageous since it reduces the mass flow of vapour which has to be absorbed to cover a certain heat demand. The difference in latent heat of vaporization between the two mixtures increases with an increasing amount of salt in solution.

It was also found by Åhlby (to be published) that for a given situation the optimum temperature gradient in the absorber is equal to the internal temperature gradient which minimizes the pressure ratio, π, across the cycle 'corrected' for the influence of the solution heat exchanger. The temperature gradient in the absorber, ΔT_{abs}, is defined as the difference between the the saturated temperature at which the absorption starts and the solution temperature in the absorber outlet, $T_3 - T_4$ in Fig. 1. Assuming there is no influence on the cycle performance by the solution loop, the optimum ΔT_{abs} is found for equal internal and external temperature gradients giving a minimum π across the cycle. If the external temperature gradients on the sink and the source sides are very different, the optimum ΔT_{abs} is found for a value in between the values of the external gradients.

As the recirculation flow varies strongly with the internal temperature gradient, the importance of the solution heat exchanger changes with the operating conditions. For small internal temperature gradients, the solution flow becomes very large and a smaller amount of the heat possible to transfer internally can actually be handled by the solution heat exchanger area available. For large ΔT_{abs}, corresponding to small solution flows, the amount of heat to transfer internally is considerably less. Thus, the influence of the solution heat exchanger is minor at large temperature gradients but very pronounced at small gradients.

The optimum ΔT_{abs} is always a compromise, the desire to minimize both the π and the heat loss in the solution loop. For large external temperature gradients, the optimum ΔT_{abs} is found close to the value giving the minimum π, while for small external temperature gradients the optimum ΔT_{abs} is found for considerably larger values than given by the external gradients. From this viewpoint the mixture H_2O/LiBr is better suited for the cycle than NH_3/H_2O since the specific heat capacity of the former is considerably lower

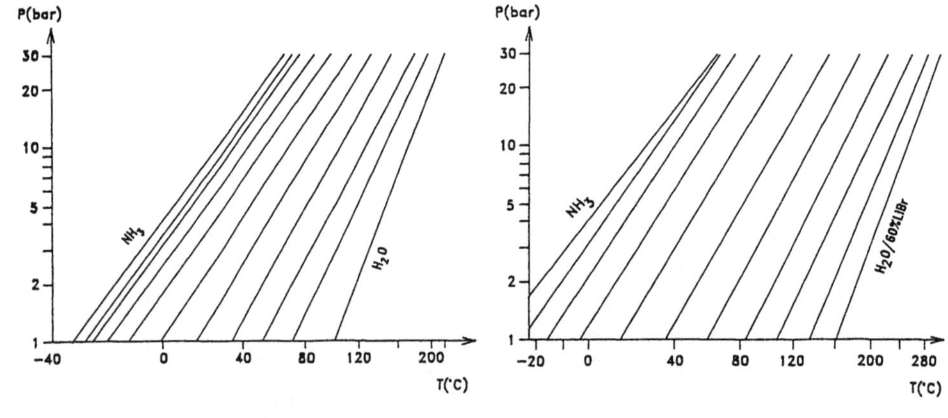

Fig. 2. Dühring diagram for (a) NH_3/H_2O and (b) NH_3/H_2O-60% LiBr.

(approximately half). Thus, the heat transferred internally in the solution heat exchanger is reduced, minimizing the heat exchanger area. The effect of a reduced specific heat is similar to the effect of a reduced solution flow, in that it reduces the amount of heat exchange. The ternary mixture NH_3/H_2O-60% LiBr exhibits a solution specific heat in between the values of the two binaries.

The lower vapour pressure of the ternary mixture, compared to NH_3/H_2O, is advantageous from the point of view of the solution pump. The pump work is proportional to the pressure difference across the cycle and is, therefore, reduced. But again there is the question whether the positive effects obtained from the reduced vapour pressure and reduced specific heat of solution (with the addition of salt) are big enough to overcome the possible increase of solution mass flow.

The compressor side of the cycle is not believed to 'suffer' from the addition of the salt. Compared to NH_3/H_2O the situation could be worse (due to the vapour pressure decrease), but then ammonia/water is extremely advantageous in this respect. Comparisons with Refrigerant 12 (Åhlby, to be published) show reductions of the compressor size by up to 30% if NH_3/H_2O is used. Thus there is a considerable margin, even though a small compressor is preferred.

The main drawback of the ternary mixture compared to the binary is its higher viscosity. Since this property is important for the heat transfer, a lower heat transfer coefficient can be expected for NH_3/H_2O-LiBr. Since no information on heat transfer is available for this ternary mixture, the difference in heat transfer between the two mixtures is ignored in the following discussion. Thus, the calculations assume equal heat transfer for NH_3/H_2O and NH_3/H_2O-60% LiBr. Other problems, such as corrosion, toxicity etc., are not considered in this study. They would have to be taken into consideration at a later stage.

3. RESULTS FROM A HIGH-TEMPERATURE CASE

3.1 General

One area of interest for heat pumps is the temperature range from 70 °C to 130 °C and higher. Today the only commonly used working fluid that can be used up to these high temperatures is water. As discussed earlier, water has the disadvantage of exhibiting very large vapour volumes at lower temperatures, and evaporation above 80 °C is usually preferred for this fluid. Other possible fluids, such as ammonia/water, have been used for high-temperature applications only to a quite limited extent so far, see for example Mucic (1990).

An application where low-pressure steam is generated (2.7 bar corresponding to 130 °C), is taken as an example to demonstrate the compression/absorption cycle with the two mixtures NH_3/H_2O and NH_3/H_2O-60% LiBr. The heat source is assumed to be hot water at 70 °C. The source is cooled by 5, 10 and 15 K respectively.

The performance of the cycle is calculated for a constant heat transfer capacity distributed according to: UA_{abs}= 200 kW/K, UA_{des}= 120 kW/K and UA_{shx}= 100 kW/K. The absorber and the desorber operate with a logarithmic mean temperature difference of approximately 5 K and the heat output from the absorber is 1 MW. The solution heat exchanger, however, operates with a temperature difference which varies strongly with the solution flow and therefore with the operating conditions.

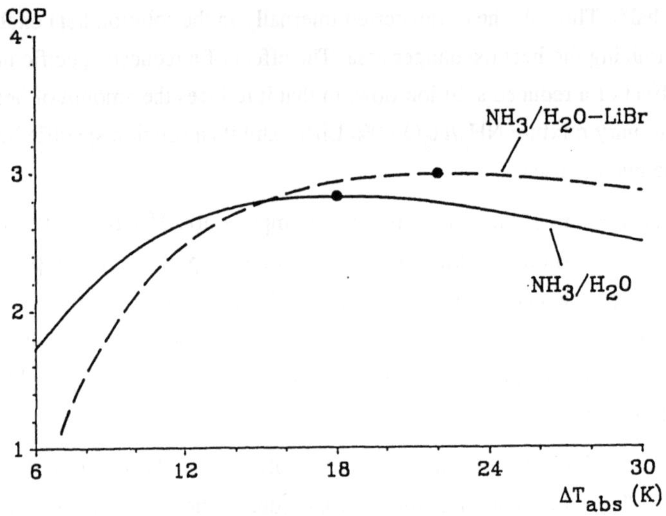

Fig. 3a. Comparison of COP for NH_3/H_2O-60% LiBr and NH_3/H_2O. T_{sink}= 130 - 130 °C, T_{source}= 70 - 65 °C.

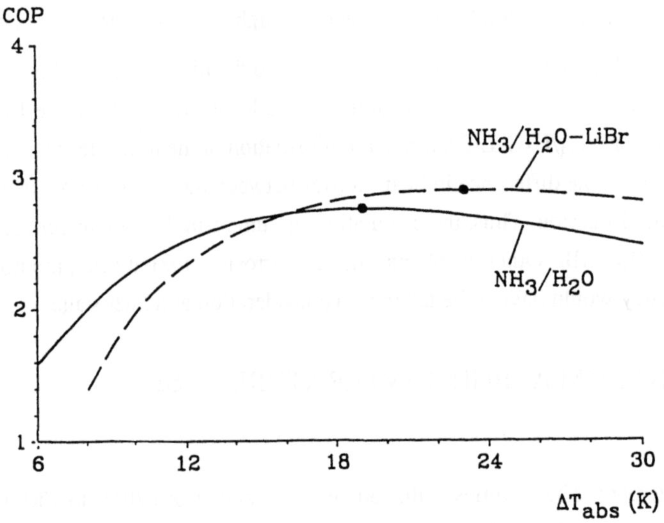

Fig. 3b. Comparison of COP for NH_3/H_2O-60% LiBr and NH_3/H_2O. T_{sink}= 130 - 130 °C, T_{source}= 70 - 60 °C.

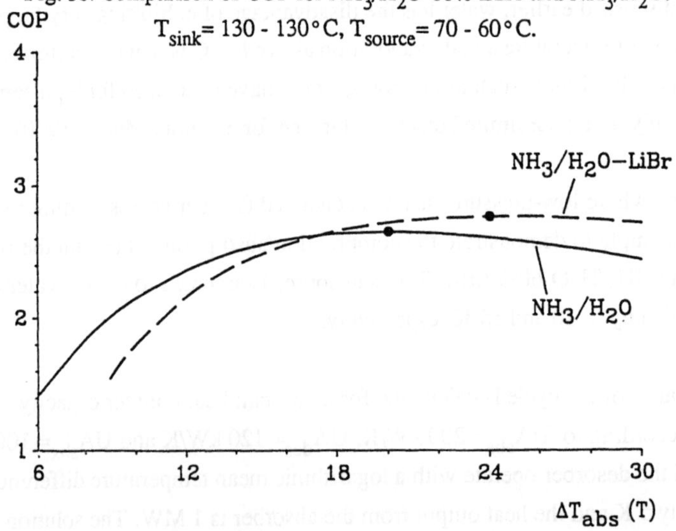

Fig. 3c. Comparison of COP for NH_3/H_2O-60% LiBr and NH_3/H_2O. T_{sink}= 130 - 130 °C, T_{source}= 70 - 55 °C.

Under the assumption of equal heat transfer coefficients for the two mixtures, the constant UA-values correspond to equal heat exchanger area for both systems. This does not mean the investment cost for the cycle is held constant throughout the calculations. The size of the compressor, expressed as the specific compressor displacement (SCD) and defined as the mass flow of vapour compressed per MJ heat output from the absorber, will still vary. The results from the calculations show that the location of the maximum COP approximately coincides with the minimum SCD, which means that the best performance is attained with the smallest (and cheapest) compressor.

The compressor considered is an oil-lubricated screw compressor. The isentropic and volumetric efficiencies, expressed as functions of the pressure ratio π, the discharge pressure P_{abs} and the type of fluid, are given in Berntsson (1986). The results from the calculations will be shown in diagrams giving the COP and the SCD as functions of the temperature gradient in the absorber ΔT_{abs}.

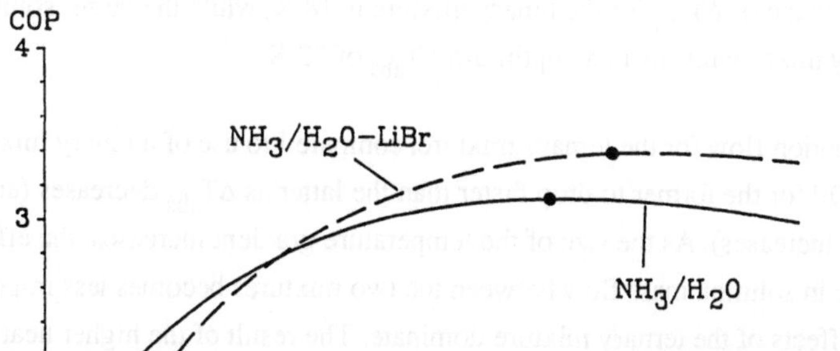

Fig. 4. Comparison of COP for NH_3/H_2O-60% LiBr and NH_3/H_2O.
T_{sink}= 110 - 130°C, T_{source}= 65 - 50°C.

In all calculations, a maximum system pressure of 25 bar is allowed. A lower pressure level at the given temperatures corresponds to a solution less rich in ammonia. It is disadvantageous, as it causes the solution flow to increase. This effect is very pronounced for the ternary mixture since the amount of salt is defined in aqueous solution. Thus, the increase in solution mass flow with an increasing water concentration is larger for the ternary mixture compared to the binary.

3.2 Results

The results from the comparison of the two mixtures are shown in Fig. 3a-3c. Figure 3a gives the COP for an external temperature gradient on the source side of 5 K. As can be seen, the binary mixture gives a higher COP than the ternary up to a COP close to the maximum value of COP for NH_3/H_2O. At larger values of ΔT_{abs} the COP of the ternary mixture exceeds that of the binary. The same behaviour is found for the other values of the source temperatures (see Fig. 3b-c). The optimal value of ΔT_{abs} (the internal temperature gradient giving the maximum COP) is moved to larger values for the ternary mixture. The location of the maximum COP for each fluid is indicated in the diagrams. The improvement achieved by using NH_3/H_2O-60% LiBr is, however, small. For the situation shown in Fig. 3a it is only 6% while, for Case 3c, it is 4%.

The variation in COP with the temperature gradient in the absorber is, as stated earlier, given by the relative influence of the variables that determine the performance of the system: the compression ratio π and the size of the solution flow, F. For the case shown in Fig. 3a, the minimum π for both the binary and the ternary fluid is found for the smallest value of ΔT_{abs}. Due to the almost isothermal external conditions, an increasing ΔT_{abs} gives a worse matching between the internal and external conditions and, thus, an increasing π. Since the influence of the solution loop is large for small values of the internal temperature gradient (large solution flows and large amounts of heat transferred internally in the solution heat exchanger) the optimum value of ΔT_{abs} is moved towards higher values. The optimum value of ΔT_{abs} for the binary mixture is 18 K, while the larger solution flow for the ternary mixture results in an optimum ΔT_{abs} of 22 K.

The larger solution flow for the ternary mixture, compared to use of a binary mixture, also causes the COP for the former to drop faster than the latter as ΔT_{abs} decreases (and the solution flow increases). As the size of the temperature gradient increases, the effect of the difference in solution mass flow between the two mixtures becomes less important, and the positive effects of the ternary mixture dominate. The result of the higher heat of vaporization for the ternary compared to the binary mixture is seen as a reduced requirement of compressor work, which is the reason for the higher COP found with the ternary mixture at large values of ΔT_{abs}.

As the external temperature gradient on the source side is increased, the relative behaviour of the cycle for the binary and the ternary mixture remains unchanged. However, the improvement gained for the ternary mixture is somewhat smaller for large temperature gradients of the source. The reduced improvement, from 6% to 4%, is caused by the steeper vapour pressure curves for the ternary mixture, giving a larger increase in π as the temperature lift across the cycle increases.

It seems as if a better situation for the ternary mixture would be for large external temperature gradients. For such a situation, the location of the minimum π across the cycle would move towards larger values of ΔT_{abs} and the negative effect of the larger recirculation flow for the ternary mixture would be reduced. Figure 4 shows a comparison of the performances for the two fluids for a case where T_{sink}= 110-130 °C and T_{source}= 65-50 °C. The temperatures are chosen to give a temperature level similar to that in Fig. 3a-c, but also an equal average temperature lift (comparable to Case 3a). In this case the improvement by using NH_3/H_2O-60% LiBr is 9% compared to NH_3/H_2O, which is still a relatively limited improvement.

The size of the compressor, expressed as the specific compressor displacement (SCD) in m^3/MJ heat output, is shown for one case in Fig. 5. Since the pressures in the cycle are approximately the same for both fluids, hardly any influence is seen on SCD. NH_3/H_2O-60% LiBr gives a larger value at small ΔT_{abs} due to the increased heat loss in the solution loop at these large solution flows. At large ΔT_{abs} the ternary mixture shows a slightly lower SCD instead, due to the higher heat of vaporization and the resulting smaller volume flow.

4. CONCLUSIONS AND DISCUSSION

The improvement gained in COP by using the ternary mixture NH_3/H_2O-60% LiBr instead of the binary NH_3/H_2O is small, only about 5% in the examples shown here. One reason for this is the unfavourable situation chosen, exhibiting isothermal heat transfer on the sink side. This, however, is a common situation in industry since there is often a demand for low-pressure steam for heating purposes. This particular situation does not allow the advantages of the ternary mixture to be fully used. A more favourable situation for the ternary mixture would be one with large external temperature gradients. Even then, however, the improvement gained in COP by using NH_3/H_2O-60% LiBr is rather small. In the example shown a 9% higher COP was obtained.

In return for the small increase in performance obtained by using the salt in the ternary mixture NH_3/H_2O-LiBr, one must accept the consequences of its higher viscosity, density and worse heat transfer coefficient.

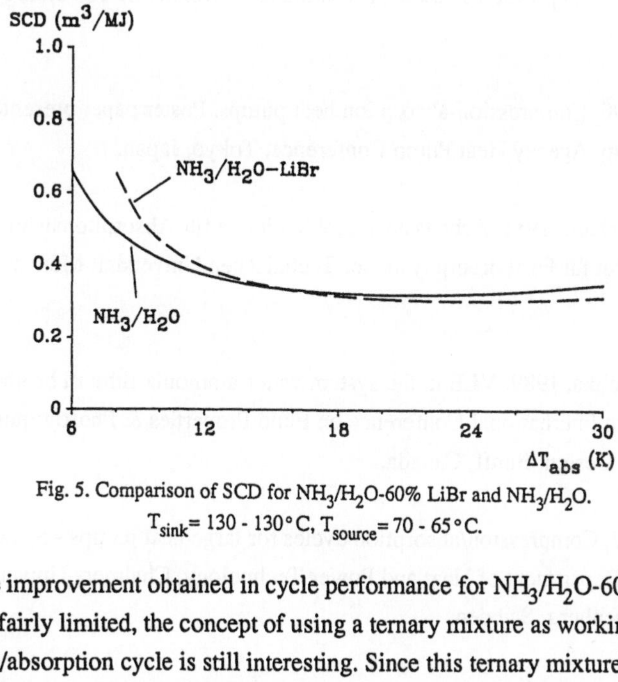

Fig. 5. Comparison of SCD for NH_3/H_2O-60% LiBr and NH_3/H_2O. T_{sink}= 130 - 130 °C, T_{source}= 70 - 65 °C.

Even though the improvement obtained in cycle performance for NH_3/H_2O-60% LiBr compared to NH_3/H_2O, is fairly limited, the concept of using a ternary mixture as working fluid for the compression/absorption cycle is still interesting. Since this ternary mixture exhibits both positive (high heat of vaporization, low solution specific heat) and negative (large recirculation flow) characteristics compared to ammonia/water, a reduction of the salt content in solution might change the picture given here. A solution with a salt content between 0 and 60% could allow the compression/absorption cycle to maximize the advantages and to minimize the disadvantages of the ternary mixture. The problem is, of course, the lack of experimental data available for such a mixture. Estimations can and will, however, be done using data for the fluids available.

It must be remembered that NH_3/H_2O-LiBr is chosen only as an example of a ternary mixture. It is possible there are other ternary mixtures that are better suited for the compression/absorption cycle. The major problem in finding these is the lack of thermodynamic data, which are necessary to calculate the cycle performance and thereby predict the potential for the fluid.

ACKNOWLEDGEMENTS

The authors would like to express their gratitude to Professor Reinhard Radermacher, University of Maryland, USA, for his guidance concerning the properties of the ternary mixture, Professor Thore Berntsson, Chalmers University of Technology, Sweden, for valuable discussions, and to the Swedish National Board for Technical Development (STU), the Swedish Council for Building Research (BFR) and the National Energy Administration (STEV) for supporting this work financially.

REFERENCES

Berntsson, Karen Munch, 1986. Nonazeotropic mixtures as working fluids in large heat pumps - heat transfer in a falling film evaporator and system simulations. Dissertation, Department of Heat and power Technology, Chalmers University of Technology, Gothenburg, Sweden.

Mucic, Vinko, 1990. Compression-absorption heat pumps. Poster paper presented at the 3rd International Energy Agency Heat Pump Conference, Tokyo, Japan.

Radermacher, Reinhard, 1981. Arbeitsstoffkombinationen für Absorptionswärmepumpen. Dissertation, Institut für Festkörperphysik der Technischen Universität München, Munich, W. Germany.

Zimmerman, Angelika, 1989. VLE in the system water-ammonia-lithium bromide. Poster paper presented at the 5th International Conference on Fluid Properties & Phase Equilibria for Chemical Process Design, Banff, Canada.

Åhlby, Lena, 1987. Compression/absorption cycles for large heat pumps - System simulations. Licentiate thesis, Department of Heat and Power Technology, Chalmers University of Technology, Gothenburg, Sweden.

Åhlby, Lena; Hodgett, David; Berntsson, Thore. The compression/absorption cycle - an optimization study. To be published.

PART II

Working fluids and CFC replacements

Chapter 7

CFC ALTERNATIVES FOR HIGH-TEMPERATURE HEAT PUMP APPLICATIONS

L Vamling, M Högberg and T Berntsson (Chalmers University of Technology, Gothenburg, Sweden)

SUMMARY

Two typical high temperature heat pump applications have been examined, production of low-pressure steam, and production of high temperature hot water. Heat pump cycle calculations have been carried out for several possible CFC alternatives. Two different methods of comparing the performance of the fluids have been used, and the importance of including the transport properties in the comparison is pointed out. It is concluded that in the near future the CFC alternatives for high temperature applications will either be flammable, or chlorine containing. Flammable fluids may give higher COPs than non-flammables and are therefore worth considering. One way to use the benefit of a flammable fluid can be to mix it with a non-flammable, in such proportions that the mixture is non-flammable. The use of mixtures can also extend the working range of fluids that otherwise will have too high condenser pressures.

NOMENCLATURE

COP	Coefficient of performance = Q_{cond} / W_{comp} (-).
M	Molecular weight (kg/kmol).
P_c	Critical pressure (Pa).
P_{cond}	Condenser pressure (Pa).
P_{evap}	Evaporator pressure (Pa).
Q_{cond}	Condenser heat output (W).
T_b	Normal boiling point (K).
T_c	Critical temperature (K).
W_{comp}	Compressor work (W).
η_{is}	Compressor isentropic efficiency (-).

Heat Pumps – Proceedings of the 4th International Conference, pp.71-82

INTRODUCTION

The expression "CFC alternatives" has, in heat pump, air conditioning and refrigeration applications, so far mostly been equated to CFC12 alternatives. The discussion on replacements for higher boiling CFCs, e. g. CFC11, has mainly been focused on their use as blowing agents. The reason for this is of course that these applications are large "CFC users". As Berntsson (1987) has pointed out, there are many high-temperature applications where heat pumps are of interest. Examples of such applications are the production of low-pressure steam from high-temperature waste water in industrial applications, and the production of high-temperature hot water. The suggested working fluid has in most of these cases been CFC114, a substance that now is out of the question due to its high ozone depletion and global warming potential.

This paper will give an overview of possible working fluids for compression heat pumps for temperatures above those where CFC12 normally is used, i. e. condensing temperatures above 70 °C, and below those where water is suitable as working fluid. The influence of differences in heat transfer and pressure drop will also be investigated. The possibility of using mixtures as working fluids in this temperature range will be discussed.

WORKING FLUID REQUIREMENTS

There are several requirements which have to be more or less fulfilled before a substance can be put to use as working fluid in a heat pump. Examples of such requirements are:

Suitable pressures
Low ozone depletion potential
Low global warming potential
Low toxicity
Non-flammability
Compatibility with construction materials and compressor oils
High COP
High capacity

The first requirement is illustrated in Fig. 1, where the possible working temperature ranges for different operating pressure requirements are given. Even if it is possible to work with evaporator pressures below atmospheric pressure, it is usually avoided due to air leakages and large volumes. The upper limit varies with compressor type but is usually between 2 and 2.5 MPa, at least for screw compressors.

At a first glance it may seem as there are a lot of substances to choose from, but, as will be shown later, the other requirements will drastically reduce the number of alternatives. Many of the fluids included in Fig. 1 are flammable, some have undesirable environmental properties and a few might be toxic. Before excluding any of the suggested fluids on these grounds it might be interesting to know whether their performance as working fluid are so much better than other alternatives that it might be worthwhile to try to cope with the undesired property. A flammable fluid with otherwise desirable properties might also be of interest to use in a mixture. For these reasons heat pump cycle calculations have been carried out for a number of the fluids in Fig. 1, and also for some mixtures. The calculations have been done for two different sets of external conditions.

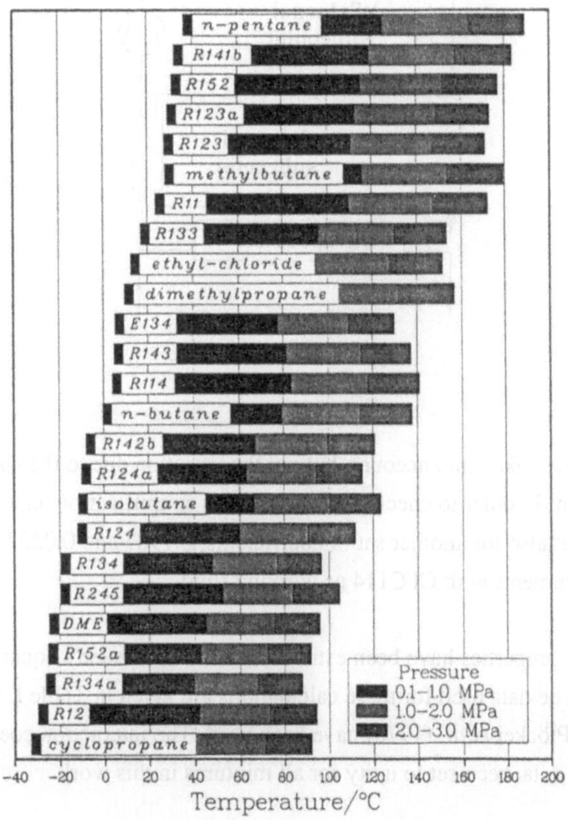

Fig. 1. Temperature ranges for some working fluids.

CONDITIONS FOR THE HEAT PUMP CYCLE CALCULATIONS

General

For simplicity a one-stage heat pump cycle has been used for all cases. The cycle is illustrated in Fig. 2. Note that both external and internal subcooling has been considered whenever applicable. The minimum approach temperature in the external subcooler was set to 3 K. The heat output from the cycle was set to be the same for all cases. All heat exchangers were calculated as being of counter current type. The isentropic efficiency of the compressor, η_{is}, was calculated as follows:

$$\eta_{is} = 0.815 - 0.02\, P_{cond}/P_{evap} \tag{1}$$

This equation is based on measurements on a twin screw compressor with adjustable built-in volume ratio. The coefficients are dependent on the choice of working fluid and compressor oil. The ones given in the equation above are typical for CFC12. Due to lack of data they have here been used for all working fluids. Hence the variation of efficiency with the

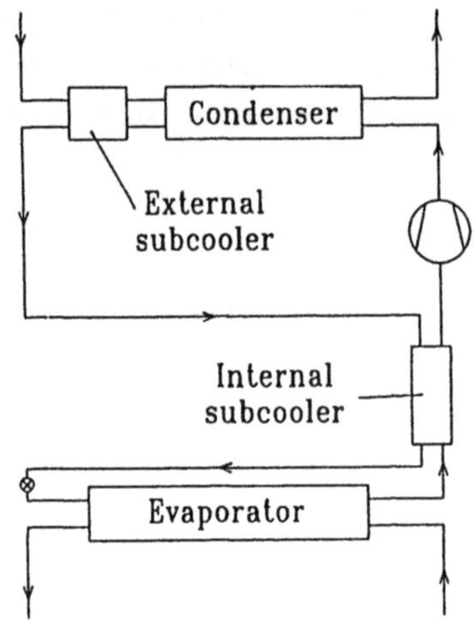

Fig. 2. The heat pump cycle.

pressure ratio has been taken into account, but not the variation due to the specific fluid-oil combination. In order to check the effect of this simplification, calculations have been carried out also for another set of coefficients, 0.7987 and 0.0227. These values are based on measurements with CFC114 as working fluid.

The thermodynamic properties have been estimated with Lee-Kesler's equation of state (Lee and Kesler, 1975). The data used for these calculations are given in Table I. For mixtures the mixing rules of Plöcker *et al.* (1978) have been used. The interaction coefficients have, due to lack of data, been set to unity for all mixtures in this work.

Since there are less data available and since prediction methods are less developed for transport properties compared to state properties, it is attractive if the order between different working fluids would only be marginally influenced by differences in transport properties. In order to investigate the importance of including the heat transfer and pressure drop in the cycle calculations two different methods of comparison have been used in this work, one simple and one more rigorous.

The Simple Method

In the simple method the minimum approach temperatures in the condenser and in the evaporator are the same for all fluids, 3 K and 3.5 K respectively. The suction gas superheat is 10 K and external subcooling is used if the the temperature difference between the bubble point temperature of the working fluid and the heat sink inlet is more than 3 K.

The Rigorous Method

In the other more rigorous method CFC114 has been used as a reference substance. For this substance the heat exchanger areas necessary to fulfill the requirements given for the simple method have been calculated. The performance of the other substances has then been calculated using these reference areas. In the evaporator, the tube length that gives the optimal balance between heat transfer and pressure drop has also been calculated. The importance of this tube length optimization has been discussed by Högberg *et al.* (1990).

Table I. Data used for calculation of thermodynamical properties.

Substance	Formula	M kg/kmol	T_b K	T_c K	P_c MPa	Data sources[a]
cyclopropane	C_3H_6	42.081	240.3	397.8	5.49	
R12	CCl_2F_2	120.914	245.2	385.01	4.129	Higashi *et al.* (1984)
R134a	CF_3CFH_2	102.030	247.0	374.20	4.056	McLinden *et al.* (1989)
R152a	CF_2HCH_3	66.051	248.2	386.44	4.52	Higashi *et al.* (1987)
DME	CH_3OCH_3	46.069	248.3	400.0	5.24	
R134	CF_2HCF_2H	102.030	253.6	392.06	4.595	Allied Signal
R245	$CF_3CF_2HCH_3$	134.047	255.7	380.1	3.14	
R124	$CClFHCF_3$	136.480	261.1	395.4	3.62	ICI
isobutane	C_4H_{10}	58.124	261.4	408.2	3.65	
R124a	$CClF_2CF_2H$	136.475	263.	399.9	3.72	
R142b	$CClF_2CH_3$	100.496	263.4	410.5	4.07	Kumagai *et al.* (1988)
n-butane	C_4H_{10}	58.124	272.7	425.2	3.8	
R114	$CClF_2CClF_2$	170.922	276.2	418.9	3.26	
R143	CF_2HCFH_2	84.041	278	(431.)	(4.23)	
E134	CF_2HOCF_2H	118.030	278.1	(418.)	(3.95)	Kopco (1990)
dimethylpropane	C_5H_{12}	72.151	282.6	433.8	3.2	
ethyl chloride	C_2ClH_5	64.515	285.5	460.4	5.27	
R133	$CClFHCF_2H$	118.490	290	(445.)	(4.08)	
R11	CCl_3F	137.368	296.9	471.2	4.41	
methyl butane	C_5H_{12}	72.151	301.0	460.4	3.39	
R123	CCl_2HCF_3	152.910	301.0	456.94	3.676	McLinden *et al.* (1989)
R123a	$CClFHCClF_2$	152.910	302.1	462.76	3.889	Allied Signal
R152	CFH_2CFH_2	66.051	304	(476.)	(4.44)	
R141b	CCl_2FCH_3	117.950	305.2	(482.6)	(4.38)	ICI
n-pentane	C_5H_{12}	72.151	309.2	469.7	3.37	

Values within parenthesis are estimated from chemical structure and normal boiling point according to the method of Joback (Reid *et al.*, 1987).

[a]When other than the property data bank in Reid *et al.* (1987)

The heat transfer in the evaporator and the condenser have been calculated by numerical integrations along the heat exchangers. For calculations of local heat transfer coefficients during the evaporation the equation of Pierre (1969) has been used. On the condenser side the equation given by Katz and Beatty (1948) has been used. The transport properties have been estimated according to Latini *et al.* (1988), and Reid *et al.* (1987).

The pressure drop in the evaporator has been calculated with the equation given by Pierre (1957). The pressure drops in the other heat exchangers have been neglected, since the differences in pressure drops between the different fluids in these heat exchangers will have a very marginal influence on the difference in COP between the fluids.

For mixtures the same heat transfer and pressure drop equations as for pure fluids have been used. The physical property data used in these equations have been taken at the actual temperature, pressure and composition. For transport properties the mixing rules given in Reid *et ai.* (1987) have been used. No additional reduction of the heat transfer coefficients for the mixtures due to mass transfer effects have been taken into account. Högberg *et al.* (1989) have discussed the influence of such a reduction, and concluded that the influence on the COP may be in the order of a few per cents.

External Conditions

Calculations have been carried out for two sets of external conditions. The first case is typical for production of hot water from waste water at a relatively high temperature. The heat source inlet and outlet temperatures in this case were set to 55 °C and 50 °C, respectively, while the heat sink inlet and outlet temperatures were 85 °C and 90 °C. The sum of the external heat transfer resistances, including the tube wall and fouling, expressed as a heat transfer coefficient referred to the heat transfer area on the working fluid side, was 1100 W/(m^2K) in the condenser and 2500 W/(m^2K) in the evaporator. In the evaporator the working fluid was assumed to be on the tube side and in the condenser on the shell side.

The second case considered is typical for production of low-pressure steam from high temperature waste water. In this case the heat source inlet and outlet temperatures were set to 75 °C and 70 °C, while the heat sink had a constant temperature of 120 °C. The "external heat transfer coefficient" in the evaporator was the same as in the first case, while the value for the condenser was set to 1560 W/(m^2K) to account for the higher value of the heat transfer coefficient for steam compared to that of water.

RESULTS AND DISCUSSION

Results for Pure Fluids

The results for the first case are given in Fig. 3, where the comparison is done using equal minimum approach temperature, and in Fig. 4, where the heat exchanger areas are the same for all fluids. The results for the second case are given in Fig. 5 and Fig. 6. In all these figures the COPs are given versus the required condenser pressure. If the COPs were given versus the capacity, the figures would have been almost the same, since the capacity is approximately proportional to the condenser pressure. Different markers have been used to notify if a substance is chlorine-containing, i. e. having a non-zero ozone depletion potential, flammable, or fully substituted. Chlorine-containing flammable fluids have been marked as flammable. Substances not belonging to any of these groups have been designated as "friendly", even if there might be uncertainties regarding their toxicity or global warming potential. The flammability data have been taken from McLinden and Didion (1987).

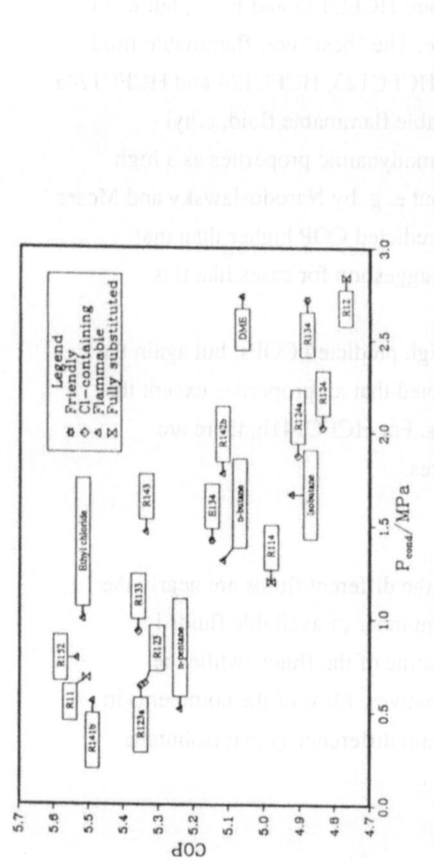

Fig. 3. Comparison of some pure working fluid with equal minimum approach temperature in condenser and evaporator - Case I

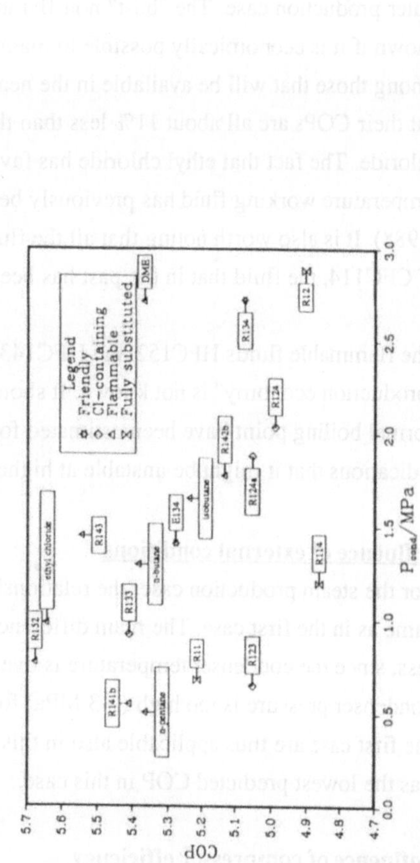

Fig. 4. Comparison of some pure working fluid with equal heat exchanger areas - Case I

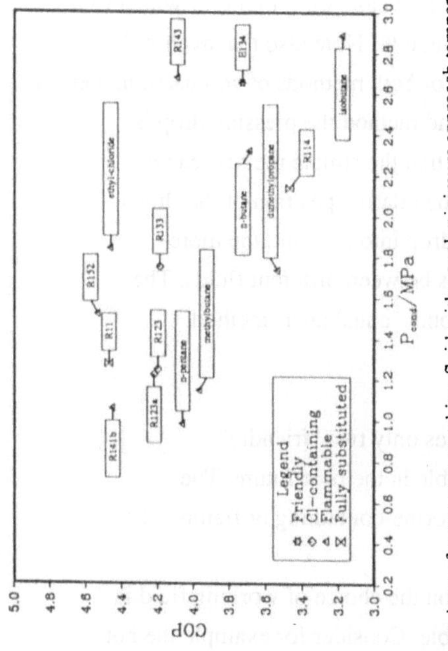

Fig. 5. Comparison of some pure working fluid with equal minimum approach temperature in condenser and evaporator - Case II

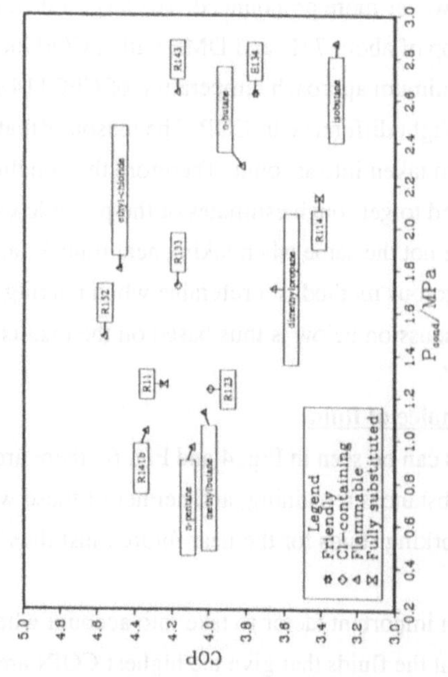

Fig. 6. Comparison of some pure working fluid with equal heat exchanger areas - Case II

Influence of heat transfer and pressure drop

By comparing the two sets of figures it can be seen that the prediction of condenser pressure is only slightly influenced by the method of comparison. The changes in COPs are, however more pronounced. The largest differences in the first case are for CFC11 with a COP drop of about 7 %, and DME with a COP increase of about 6 %. Note also that even if the minimum approach temperature of CFC114 is the same for both methods of comparison, there is a slight difference in COP. The reason is that in the second method the pressure drop is also taken into account. Therefore the conclusion is then that the simple method can be used to get rough estimates of the possible COP, but as the relative position of the fluids are not the same when taking heat transfer and pressure drop into account, the more rigorous method is preferable when making comparisons between different fluids. The discussion below is thus based on the results for this second "equal area" method.

Choice of fluid

As can be seen in Fig. 4 and Fig. 6., there are in both cases only two "friendly" substances remaining, and neither of these will be available in the near future. The working fluids for the near future must thus be either chlorine-containing or flammable.

An important factor to take into account when deciding on the choice of working fluid is that the fluids that give the highest COPs are all flammable. Consider for example the hot water production case. The "best" non-flammable fluids are HCFC133 and E134, but it is not known if it is economically possible to manufacture these. The "best" non-flammable fluid among those that will be available in the near future are HCFC123, HCFC124 and HCFC124a, but their COPs are all about 11% less than the best available flammable fluid, ethyl chloride. The fact that ethyl chloride has favourable thermodynamic properties as a high temperature working fluid has previously been pointed out e. g. by Narodoslawsky and Moser (1988). It is also worth noting that all the fluids have a predicted COP higher than that of CFC114, the fluid that in the past has been the usual suggestion for cases like this.

The flammable fluids HFC152 and HFC143 have also high predicted COPs, but again the "production economy" is not known. It should also be noted that all properties except the normal boiling point have been estimated for these fluids. For HCFC141b, there are indications that it might be unstable at higher temperatures.

Influence of external conditions

For the steam production case, the relationship between the different fluids are nearly the same as in the first case. The main difference is that the number of available fluids is less, since the condenser temperature is overcritical for some of the fluids, while the condenser pressure is too high (> 3 MPa) for other alternatives. Most of the comments in the first case are thus applicable also in this case. The main difference is that isobutane has the lowest predicted COP in this case.

Influence of compressor efficiency

One important question is to what extent the uncertainty regarding compressor efficiency could influence the results presented above. The use of CFC114-coefficients in Equation 1 resulted in efficiency-drops around 3.4%, which in turn caused COP-drops between 2 and 3%.

This is a good indication on the uncertainty since CFC12 is in the middle range and CFC114 in the low range of known compressor efficiencies. The conclusion is then that the relative order between individual substances may change, but the main conclusions above are still valid.

Results for Mixtures

One of the conclusions from the discussion above is that the number of available and acceptable pure compounds is rather low, especially if non-flammability is one of the desired properties. One way of extending the working range of the fluids is to mix them. In this way the capacity can be held at an acceptable level without getting too high a condenser pressure, and as already mentioned the COPs can be improved for the available non-flammable fluids by mixing them with a flammable fluid with favourable properties, in such proportions that the mixture is non-flammable.

Three examples of mixtures that can be of interest are HCFC123/HCFC124a, HCFC123/HCFC142b, and HCFC123/C_2H_5Cl. The performances of these mixtures are shown in Fig. 7 and Fig. 8 for the two applications of hot water production and low-pressure steam generation, respectively. The method used for comparisons is the one with equal heat exchanger areas. As can be seen in the figures the COP increases with the content of ethyl chloride in the mixture of HCFC123/C_2H_5Cl. Both the mixtures HCFC123/HCFC124a and HCFC123/HCFC142b increase the capacity at the expense of a higher condenser pressure.

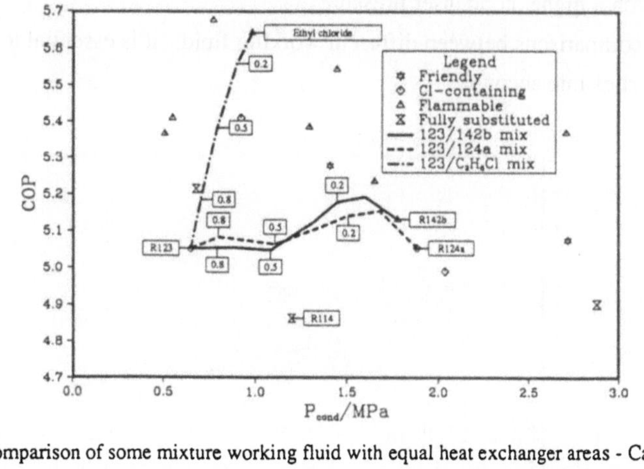

Fig. 7. Comparison of some mixture working fluid with equal heat exchanger areas - Case I
The numbers along the mixture lines represents the mole fraction of HCFC123

Our estimates of the maximum temperature change for the mixtures during evaporation at 1 bar are 12 K, 12 K, and 2 K for HCFC123/HCFC124a, HCFC123/HCFC142b, and HCFC123/C_2H_5Cl respectively. Henceforth, the temperature changes during evaporation and condensation are called "internal gradients". According to Munch Berntsson (1986) these temperature changes are favorable for the COP of the mixtures when the heat sink and heat source have similar temperature profiles, or when the temperature changes are larger on the water side than on the working fluid side.

In both Fig. 7 and Fig. 8 the shape of the COP/pressure curve for HCFC123/C_2H_5Cl is quite linear. This mixture has very small internal gradients, hence the mixture does not give any performance improvement compared to the pure fluids in both cases. The mixtures HCFC123/HCFC124a and HCFC123/HCFC142b have larger internal gradients and in the hot water production application with a properly chosen composition it is possible to reach a slightly higher COP than for the two pure fluids. On the other hand, in the low-pressure steam generation case the internal gradient in the condenser is a disadvantage for these mixtures and no COP improvements can be observed.

The proportion of HCFC123 in the mixtures HCFC123/C_2H_5Cl, and HCFC123/HCFC142b necessary to make them non-flammable are not known to the authors.

CONCLUSIONS

For the two high temperature cases considered the following conclusions can be drawn:

- The alternative working fluids available in the near future are either flammable, or chlorine containing.
- Flammable fluids may give markedly higher COPs than non-flammables, and are thus worth serious consideration.
- The COP of a non-flammable fluid can be increased by mixing it with a flammable fluid, without getting a flammable mixture. The capacity of a fluid can be increased by adding another fluid with a higher condenser pressure.
- When making comparisons between different working fluids, it is essential to take transport properties into account.

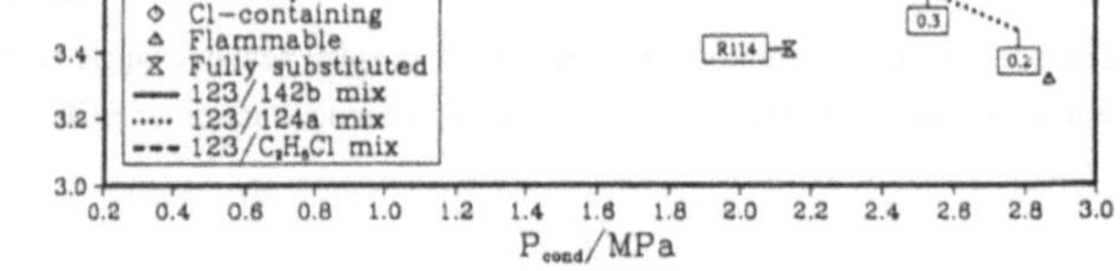

Fig. 8. Comparison of some mixture working fluid with equal heat exchanger areas - Case II
The numbers along the mixture lines represents the mole fraction of HCFC123

ACKNOWLEDGEMENTS

The authors would like to express their gratitude to the Swedish National Board for Technical Development (STU), the Swedish Council for Building Research (BFR) and the National Energy Administration (STEV) for supporting this work financially.

REFERENCES

Berntsson, T. (1987). Future prospects for industrial heat pumps in Europe in *Heat Pumps: Prospects in Heat Pump Technology and Marketing (Proceedings of the 1987 IEA Heat Pump Conference)*, K. H. Zimmerman (Ed.), Lewis Publ. Inc. Chelsea, Michigan, USA.

Higashi, Y., S. Okazaki, Y. Tskaishi, M. Uematsu, and K. Watanabe (1984). *J. Chem. Eng. Data*, 29(1), 1984, 31.

Higashi, Y., M. Ashizawa, Y. Kabata, T. Majima, M. Uematsu, and K. Watanabe (1987). Measurements of vapor pressure, vapor-liquid coexistence curve and critical parameters of refrigerant 152a, *JSME Int. J.*, 30(265), 1106.

Högberg, M., L. Vamling, and T. Berntsson (1990). Nonazeotropic mixtures as substitutes for today's CFC fluids, *Proceedings, 3rd IEA Heat Pump Conference, Tokyo.*

Naradoslawsky, N. and F. Moser (1988). New compression media as replacements for CFCs, *Int. J. Refr.*, 11(4), 264.

Katz, D.L. and K.O. Beatty (1948). Condensation of vapors on outside of finned tubes. *Chemical Engineering Progress*, 44(1), 55-70.

Kopco, W. L. (1990). Beyond CFCs: Extending the search for new refrigerants, *Int. J. Refr.*, 13(2), 79-85.

Kumagai, K., N. Yada, and K. Watanabe (1988). *Proc. 29th High Pressure Conference of Japan*, 118.

Latini, G., F. Marcotullio, P. Pierpaoli, and A. Ponticiello (1988). Dynamic viscosity of refrigerants and refrigerant mixtures: A prediction method. *Preprints of the 1988 Purdue IIR Conference*, 50-56.

Lee, B.I. and M.G. Kesler (1975). *AIChE J.*, 21(3), 510-527.

McLinden, M.O. and D.A. Didion (1987). Quest for alternatives, *ASHRAE J.*, 32.

McLinden, M.O., J.S. Gallagher, L.A. Weber, G. Morrison, D. Ward, A.R.H. Goodwin, M.R. Moldover, J.W. Schmidt, H.B. Chae, T.J. Bruno, J.F. Ely, and M.L. Huber (1989). Measurement and formulation of the thermodynamic properties of refrigerants 134a and 123, *ASHRAE Trans.*, 95(2).

Munch Berntsson, K. (1986). Nonazeotropic mixtures as working fluid mixtures and high temperature working fluids in large heat pumps - Heat transfer in a falling film evaporator and system simulations. *Dissertation*, Dept. of Heat and Power Technology, Chalmers University of Technology, Gothenburg, Sweden.

Pierre, B. (1957). Boiling flow resistance for refrigerants (in Swedish). *Kyltekn.Tidskr.*, 16(6), 225-238.

Pierre, B. (1969). Boiling heat transfer for refrigerants in horizontal tubes (in Swedish). *Kyltekn.Tidskr.*, 28(5), 128-137.

Plöcker, U., H. Knapp,and J. Prausnitz (1978). *Ind.Eng.Chem.Proc.Des.Dev.*, 17(3), 324-332.

Reid, R. C., J. M. Prausnitz, and B. E. Poling, (1987). *The properties of gases and liquids*. Fourth Ed, McGraw-Hill, New York.

Varma, H.K., C.P. Sharma, and M.P. Mishra, (1979). Heat transfer coefficients during forced convection evaporation of R12 and R22 mixtures in annular flow regime. XVth International Congress of Refrigeration, Venice, Italy. *Proceedings* Vol. II.

Chapter 8

PREDICTION OF THERMODYNAMIC PROPERTIES OF CFC MIXTURES USING THE CCOR EQUATION OF STATE

R E Low and C L Pritchard (University of Edinburgh, UK)

Abstract

The paper discusses the prediction of thermodynamic properties of refrigerants and their mixtures by use of the equation of state technique. The application of this technique to a generalised cubic equation of state, the *Cubic Chain-of-Rotators* (CCOR) equation is described. Some results of the predictions of this equation for various fluids are presented in graphical form; the predicted data are compared with experimental data and with the predictions of a more complex equation, the *Carnahan-Starling-DeSantis* equation. The facility of use of this technique for prediction of VLE data for mixtures of CFC replacements is indicated and estimation is made of the likely errors in such predicted data.

1 INTRODUCTION

The use of binary, zeotropic mixtures of chlorofluorocarbons (CFCs) as working fluids for vapour compression heat pump cycles has received increasing attention in recent years. The two main motives for use of such a mixture are: an enhancement of the COP of the cycle by utilisation of "gliding" temperature differences in the heat exchangers, and a potential for smooth variation of the thermodynamic properties of the working fluid. The latter enables a variation in the volumetric capacity (and hence a means of control over the heat output) without close control of compressor speed. It often may be the case that the flexibility of operation afforded to a heat pump by such an arrangement is of more value than the actual increase in COP. With this in mind, it should be evident that the ability to calculate the variation of fluid composition (and hence capacity) with process temperatures and pressures is a perquisite for the design or analysis of such heat pump cycles.

The capacity of a compression cycle is directly related to the suction pressure at the compressor inlet. This, in turn, is closely related to the dewpoint pressure in the evaporator. For a given evaporator dewpoint temperature, with a pure fluid this is constant. For a binary mixture, on the other hand, the pressure will vary smoothly between the vapour pressures of the two components, yielding a variation in volumetric capacity such as that shown in Figure 1. This diagram is, in

Heat Pumps – Proceedings of the 4th International Conference, pp.83-100

fact, the result of specific calculation of a novel compression cycle described elsewhere [13, 14] but the qualitative result is valid for any compression cycle. It may be seen that the variation in capacity is rather sensitive to composition, if there is an appreciable difference in the volatilities of the two components. (The heat pump cycle under investigation at Edinburgh University uses this property as a basis for self-regulation of the capacity to match the load.) In terms of cycle design, this implies that a reasonably accurate method for calculation of the vapour-liquid equilibrium (VLE) of the working fluid is desirable. It may be argued at this point, that since the evaporator dew pressure is the key to the capacity, a sophisticated technique is not required: the vapour density of most CFCs at typical source temperatures may well be so low as to render Raoult's Law valid. This is true; however,at condenser pressures, then Raoult's Law will not be adequate. In the case of the Edinburgh project, the degree of separation of the components of the working fluid by use of partial condensation is important to the performance of the system. For this reason it was decided to develop computer software capable of yielding VLE and thermophysical data for CFCs and their binary mixtures. This paper describes some results of the work as of December 1989.

A major influence on the choice of method was the need to combine ease of use in a microcomputer-based environment with minimal computational overhead: the intention was to use the method as part of a pilot-plant monitoring system. Furthermore, the method would be required to deal,

FIGURE 1: Volumetric Capacity vs composition

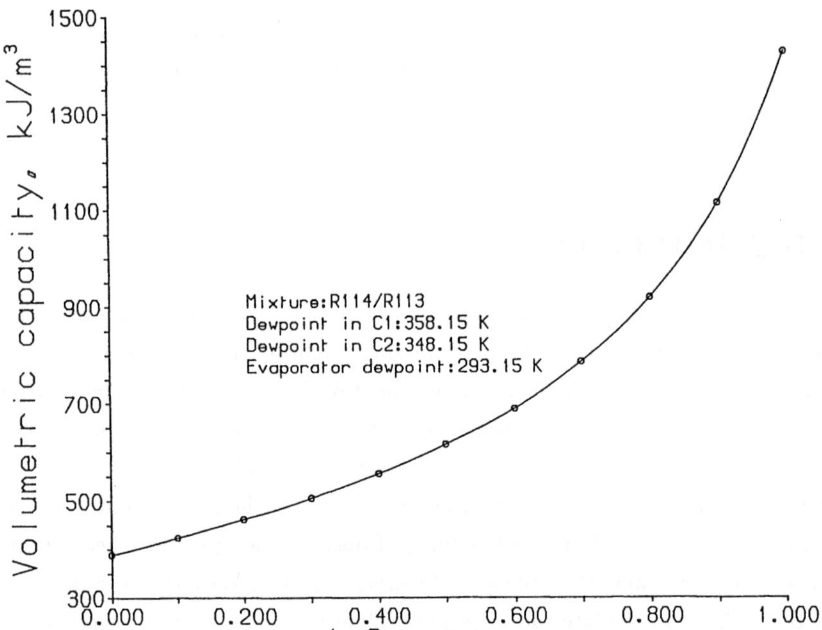

not only with the well-known refrigerants, for which accurate property correlations already exist, but also with their projected "ozone-friendly" replacements (for which relatively few experimental data existed at the commencement of the work). With this in mind, it was decided to use a predictive equation of state to represent both liquid and vapour phases. The equation chosen, as discussed below and elsewhere by Low [5] was the "Cubic Chain Of Rotators" (CCOR) equation. [4, 2, 3] This equation has previously been applied to the prediction of properties of pure CFCs and HCFCs; the work described in this paper is the extension to mixtures of CFCs. Some preliminary results of this work are compared with experimental data as an indication of the utility of this equation for preliminary prediction of refrigerant mixture VLE.

2 APPLICATION OF AN EoS TO PURE FLUIDS

The use of an equation of state as a source of thermodynamic data has received much attention, particularly for systems (e.g. hydrocarbons) in which the liquid phase exhibits near-ideal behaviour. By use of an equation capable of describing both liquid and vapour phases, it is in principle possible to generate all the data required for study of a binary mixture as a working fluid. Such an equation is often a straightforward way of dealing with a mixture at temperatures above the critical temperature of one component, where the mixture is still stable as a liquid. (The use of activity-coefficient techniques in such cases may give rise to ambiguities over the definition of standard states for the supercritical component.) A principal disadvantage is the inevitable trade-off between accuracy, speed and ease of solution.

The well-known, cubic-form equations, such as the Soave-Redlich-Kwong or Peng-Robinson equations, give good VLE prediction but are not very good predictors of the volumetric properties of the liquid phase. They have the advantage that one need only know two or three parameters- normally the critical properties and acentric factor- to use them. More complex equations tend to require determination of more parameters from experimental data for each component. This causes three problems: a need for data analysis (i.e. access to accurate experimental data); a larger computing overhead in calculation of the parameters for the mixture, and a need for iterative solution of the equation for volume, given pressure and temperature.

In view of these problems, two equations of state were selected for use in the project. The first of these was the Carnahan-Starling-DeSantis (CSD) equation as used by McLinden & Morrison [7] . This equation, a fifth order polynomial in volume, has been shown by Morrison [9] to be accurate and reliable for use with binary CFC mixtures. However, it requires six empirical constants to be fitted to data, and so was adjudged unsuitable for direct prediction of the properties of "new" refrigerants, since insufficient PVT data were available for the fluids of interest. Nonetheless, the CSD equation was implemented as a reference technique and source of data on conventional refrigerants and their mixtures.

The second equation chosen was the CCOR equation mentioned above. It requires knowledge only of the critial temperature, critical pressure and acentric factor to be applied to a substance; although regression on experimental data may be used to obtain optimal parameter sets. This equation was chosen in preference to other simple equations, since the source literature indicated superior performance for prediction of VLE and liquid phase properties. [4, 2, 3] The data necessary for use of this equation could be obtained *a priori*, using group contribution techniques. Results were reported [5] for CFCs using the methods of Lydersen and Vetere [11]. More accurate group contribution techniques for refrigerants are now available [16] but were not used in the original work.

An equation of state is typically a relation defining pressure, or compression factor, as an explicit function of volume and temperature. The dependence on composition is implicit, through the calculation of mixture parameters. This expression is then manipulated, using well-known exact thermodynamic relations (see e.g. [11]) to yield a set of relations defining fugacity, enthalpy and entropy as functions of volume, temperature and composition. These results are summarised in Appendix 1, with the resulting formulae for the CCOR equation. These non-iterative expressions, together with the quadratic or cubic polynomial describing the temperature variation of ideal gas heat capacity, may be thought of as the building blocks for construction of a thermodynamic database. In order to use these relations, it is however necessary to devise efficient strategies for the following tasks:

1. Calculation of the parameters of the equation of state.
2. Calculation of temperature and composition derivatives of these parameters.
3. Implementation of parameter mixing rules.
4. Solution of the equation for volume given temperature, pressure and composition.
5. Determination of vapour pressure of a pure component.
6. Solution of the dew or bubble point problem.
7. Solution of a generalised flash, given a known feed.

The first five tasks may be regarded as fundamental, rate-determining steps, in that they must be repeated many times in any more complex calculation (such as a flash calculation). A major reason for the popularity of cubic equations is the supposition that solution of the equation for volume is the most time-consuming step of a complex calculation. In fact, it has been shown [6, 17] that for mixtures, the calculation of the equation of state parameters and their derivatives may require as much, or more, time than the equation solution. The mixing rules chosen for this task have a significant effect on the time requirement: in this work the simplest rules, namely the van der Waals' rules (Appendix 1) were used. These are not entirely satisfactory since the resulting parameters must be corrected by empirical interaction constants. However, their computing overhead is small, compared to more complex rules. (Interaction constants will be discussed later.) Determination of the vapour pressure of a pure component (or, conversely, determination of the saturation temperature for a given pressure) is accomplished by determining the conditions under which the fugacities of the liquid an vapour phases are equal. This may be expressed as:

$$\log\,\phi^v(v^v,T) - \log\,\phi^l(v^l,T) = 0 \tag{2}$$

The vapour and liquid phase volumes are not, however, known, so that in fact the problem is 3-dimensional: solve

$$\begin{aligned} P(v^v,T) &= P^* && (3)\\ P(v^l,T) &= P^* && (4)\\ \log\,\phi^v(v^v,T) &= \log\,\phi^l(v^l,T) && (5) \end{aligned}$$

A nested calculation strategy was chosen to solve this problem using the CCOR equation. This avoids the problems which may sometimes bedevil multi-dimensional iteration and takes advantage of the availability of an analytic solution of the equation of state, which yields both phase volumes at once. The basic strategy is:

1. A guess is made at the vapour pressure.
2. Using the chosen pressure, the equation of state is rearranged into its standard polynomial form. If it has only one real root, this is flagged and the procedure is terminated. Otherwise, the equation is solved: the largest root is taken as vapour phase, the smallest as liquid phase volume.
3. The logarithms of the fugacity coefficients are found for both phases; the value of the function 2 above is found.

4. The function value from step (3) is tested against a given tolerance: if it is satisfactory this is signalled and the procedure and the procedure broken. Otherwise, the pressure is updated and flow returns to step (2) above.

This algorithm resembles that used by Morrison & McLinden for use with the CSD equation of state: it was, however, possible to make a number of simplifications, principally in the initial choice of pressure and in the pressure updating method.

The starting guess for pressure used by the author's procedures was found by assuming that the vapour pressure curve between the normal boiling point and the critical point could be approximated by the relation

$$\log P^* = A - B/T \tag{6}$$

Simple algebra yields the values of 'A' and 'B' in 6, which may then be used to find a starting guess. Morrison's more robust method finds the turning points of the isotherm curve and uses these as bounds on pressure and the phase volumes. This is a highly desirable method for solution of a non-cubic equation of state: the simpler method described above was, however, found to be adequate for all of the compounds studied.

FIGURE 2: Vapour pressures of CFCs

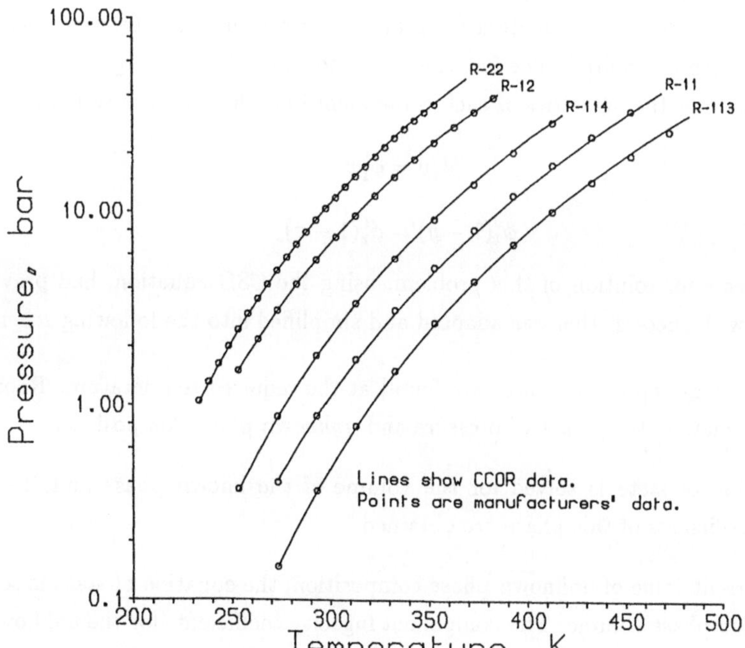

The iteration scheme described by Morrison uses a combination of the secant method with a checking mechanism to prevent divergence. This uses the known pressure bounds to obtain a regula-falsi correction, should the value of pressure update found by the secant method step outside the bounds. The methods tested in this work were: the secant method; the Newton-Raphson method with numerical derivatives, and the 'Improved Memory Method' of [15]. The technique used in the work described in [5] was the Newton-Raphson technique: this converged with no problems to a tolerance of 10^{-6} in 3-5 iterations. When the original programs were rewritten for microcomputer use, the secant method was used (for conformity with the parallel set of routines using the CSD equation). A third method, the IMM technique, is under evaluation for use in these calculations. This method for solution of a non-linear equation fits a Lagrange interpolation polynomial through the known points obtained from previous iterations. It uses more temporary storage than other methods, but does not require derivative evaluations and has been shown to require significantly less function evaluations than the other techniques. In the case of calculations such as the vapour pressure determination, where the iteration function (the difference in fugacity between phases) requires several computational steps, this appears to yield significant savings in CPU time.

The results of this method have been discussed in detail in [5], both for common CFCs and their mooted replacements. A graphical illustration is Figure 2, a plot of vapour pressure against temperature for several compounds. The prediction is in general good. The equation tends to underpredict saturated vapour volumes but the likely errors in the temperature and pressure ranges of interest for heat pumping were considered to be acceptable. The results of this work were sufficiently promising for the equation to be extended to binary mixtures.

3 APPLICATION TO MIXTURE PROPERTIES

The calculation analagous to the pure component vapour pressure determination is the isothermal dew or bubble pressure determination. If a binary mixture is used, this problem has four unknowns: the pressure and phase volumes (as before), together with the composition of the unknown phase. The method chosen to attack this problem was an extension of the nested iteration strategy described above. In addition to an iterative solution for pressure, an iterative solution for the unknown composition is required. The functions used to determine convergence for both of these loops must ensure that the fugacities of each component in each phase are equal, i.e.

$$\phi_1^v y = \phi_1^l x \tag{7}$$

$$\phi_2^v (1-y) = \phi_2^l (1-x) \tag{8}$$

An iterative scheme for solution of this problem, using the CSD equation, had previously been coded and used with success: this was adapted and simplified into the following scheme:

1. Pure component vapour pressures are found at the required temperature. Raoult's Law is used to estimate initial values of pressure and unknown phase composition.

2. The equation of state is solved for the volume of the known phase and the component fugacity coefficients of this phase are obtained.

3. For the current value of unknown phase composition, the equation of state is solved to find the unknown phase volume. The component fugacity coefficients for the unknown phase are obtained.

4. The composition iteration is tested for convergence: if it has converged, proceed, otherwise update composition and return to (3).

5. The pressure iteration is tested for convergence: if it has converged, break the iteration, otherwise update pressure and return to (2).

This iteration scheme was used to calculate all the data presented below but was found to be lacking in the necessary robustness under certain conditions. The fundamental stumbling blocks to success of this scheme are:

- It is not possible to make an initial guess of pressure if the temperature is above the critical temperature of the more volatile component.

- Raoult's Law was found to give poor initial estimates of pressure and composition at high pressures.
- As the critical region is approached, it is possible to obtain combinations of pressure and composition which do not give a valid volume root for one or other of the phases.

In fact, it may be inferred that the key to the problem is in the choice and control of pressure estimates: a better method should obtain bounds on pressure before commencement of the iteration. Such problems are not unique to the CCOR equation: other cubic equations exhibit the same behaviour. Various schemes for dealing with these problems have been suggested [1, 17, 12]. A secondary problem, unrelated to the initial pressure estimation, lies in the calculation of the equation of state parameters for each phase. These are obtained by combination of the parameters for the pure components at the same temperature, using the mixing rules described in Appendix 1. The expressions for the parameters 'a' and 'c' involve two empirical interaction constants, k_a and k_c, which are intended to account for the interactions between unlike molecules. In the absence of experimental data, it is simplest to set these both to zero, given that *a priori* calculation of these constants may not be accomplished with any predictable accuracy (Pesuit [10]). It was found that the size and shape of the two-phase envelope was quite sensitive to these constants, especially as conditions approached the critical region. The basic algorithm described above failed in some cases where non-zero interaction constants were used, and indeed failed in some cases if the interaction constants were not assigned some arbitrary values. Although the near-critical region should be of only limited interest for heat pump applications, this is a reminder that the choice of interaction constants will affect the prediction quite markedly. The quantitative effect of k_a and k_c is demonstrated below.

4 IMPLEMENTATION OF THE ALGORITHMS

The original implementations of the thermodynamic properties packages using the CCOR and CSD equations of state were coded in the programming language imp80 (a Pascal variant), to run on the Edinburgh University mainframe computer. The programs were rewritten in the C programming language for use on an IBM PC system, using a Borland "Turbo C" compiler. In both cases double-precision arithmetic was used throughout. C was chosen as the language for PC use as it offers a high degree of compatibility between the PC and Sun UNIX compiler language dialects: it may also be used more easily than FORTRAN to produce graphical output on a PC system.

The convergence tolerance used was normally 10^{-7} on a relative error. The time taken to solve the pure component vapour pressure problem was on average 0.9 seconds, using an unexpanded 10 MHz PC-XT compatible with no math coprocessor. The average number of iterations was 3. The time taken for the binary dewpoint problem was on average 3.5 seconds on the same machine. The average iteration count for this problem was 5.

5 RESULTS: VLE PREDICTION FOR MIXTURES

This section presents the predictions obtained by use of the CCOR equation for the R22/R11 system, as a typical example of a binary refrigerant system. The resulting data are compared with experimental data of [8]. The effect of the two interaction constants k_a and k_c is demonstrated; it is shown that an optimum combination of k_a and k_c must exist for any given temperature. The predictions of the CSD equation of state for the same data sets are compared with those of the CCOR equation, to establish whether one equation yields significantly better predictions when all interaction constants are set to zero.

FIGURE 3: R22/R11 isothermal bubble pressures

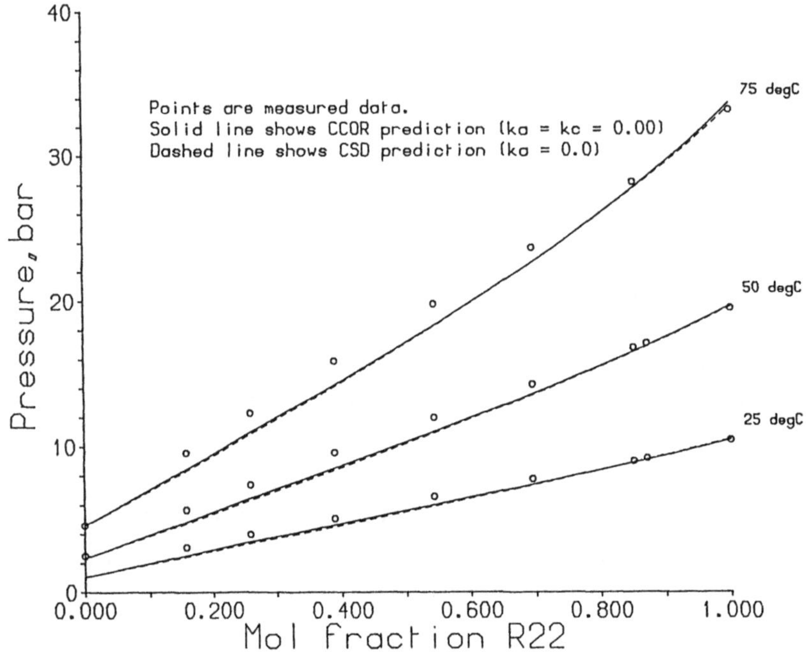

5.1 Bubble point pressure prediction

Figure 3 shows the variation of bubble pressure with composition for the R22/R11 system at temperatures of 25, 50 and 75 °C. Three sets of data are shown for each temperature: a set of experimental data, shown as points; a set of predicted data, obtained from the CCOR equation with both interaction constants set to zero, and a set of data predicted from the CSD equation with its interaction constant set to zero. The two equations yield sets of results which are virtually indistinguishable from one another at each temperature and show a systematic underprediction of bubble pressure. (Both equations do however represent the pure component vapour pressures well.) The conclusion to be drawn is that the simpler equation yields results which are *no worse* than those given by the more complex equation, if no attempt is made to evaluate interaction constants.

Figures 4 and 5 show the effect of k_a and k_c respectively on the prediction of bubble pressure by the CCOR equation. The effect of increasing k_a is to elevate the predicted pressure, while increasing k_c depresses the predicted pressure. It is evident, then, that a positive non-zero value of k_a will improve the prediction of pressure.

5.2 Saturated liquid volume prediction.

Figure 6 shows the variation of saturated liquid volume at the bubble point with composition, for the R22/R11 system. As in the comparison of bubble pressure, three sets of data are presented for each temperature. There is a noticeable difference in the shapes of curve predicted by the CCOR and CSD equations: the CSD prediction follows the curve of experimental data and exhibits

FIGURE 4: R22/R11 isothermal bubble pressures

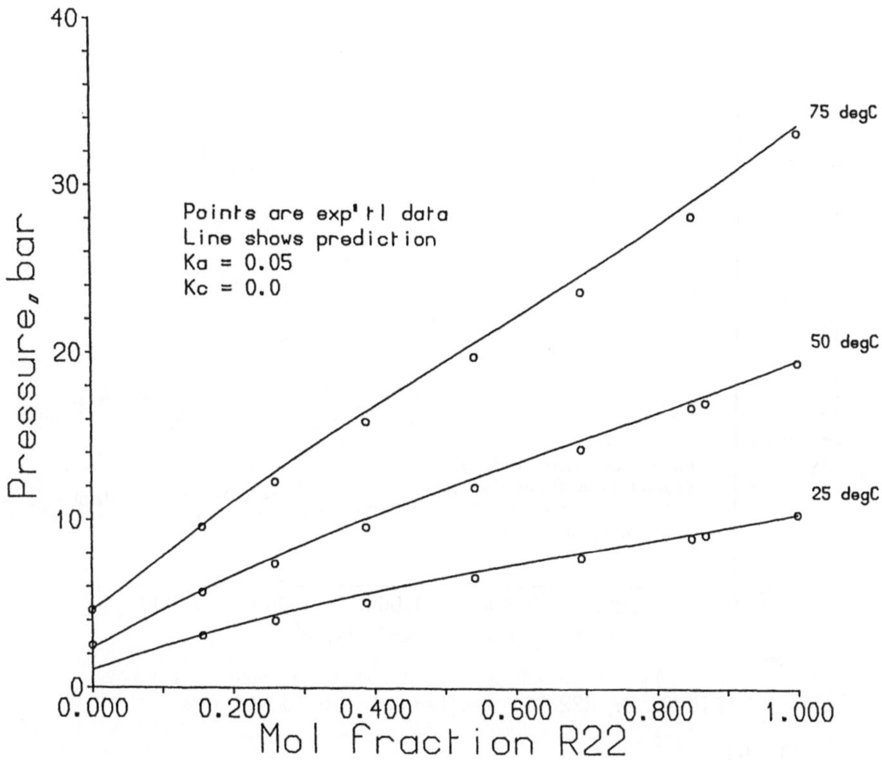

FIGURE 5: R22/R11 isothermal bubble pressures

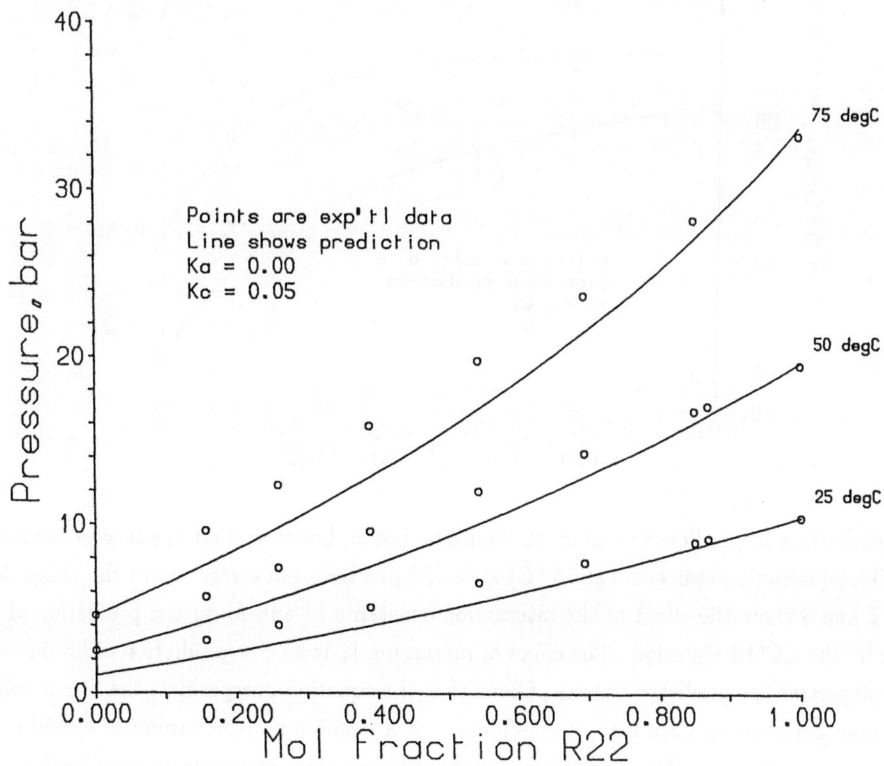

FIGURE 6: R22/R11 isothermal liquid volumes

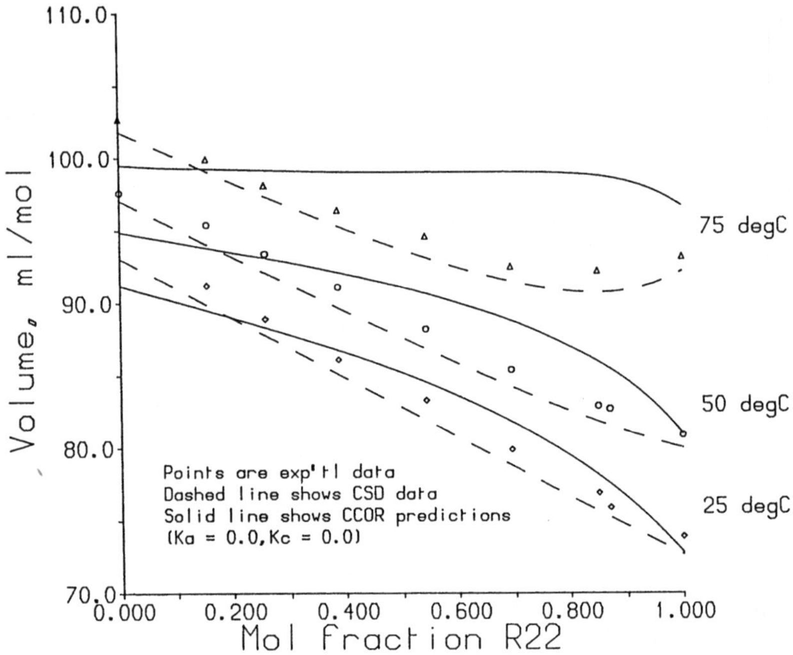

FIGURE 7: R22/R11 isothermal liquid volumes

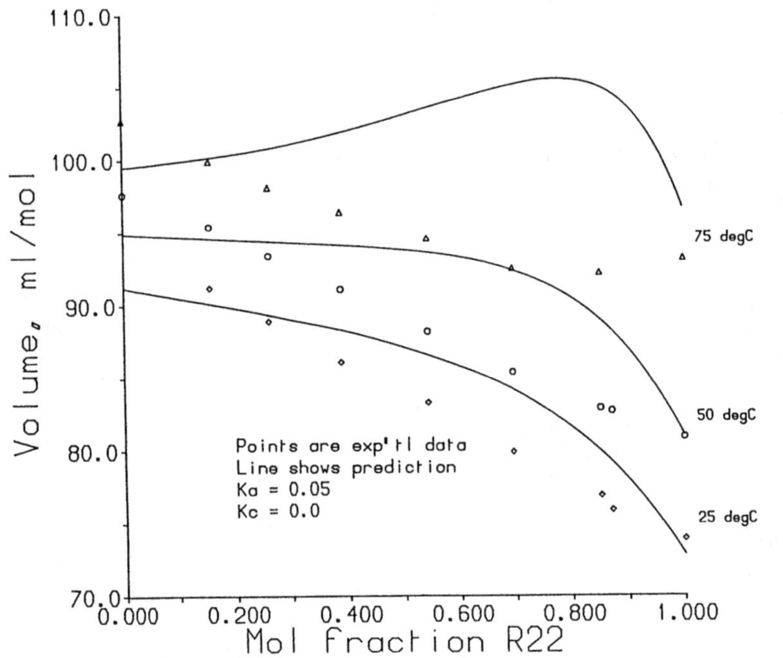

smaller deviations in predicted volume. It should be noted, however, that the largest deviation in the CCOR equation's prediction (at 75 °C) is ca. 7%, while mean errors are in the range 3-5%. Figures 7 and 8 show the effect of the interaction constants k_a and k_c on the prediction of liquid volumes by the CCOR equation. The effect of increasing k_a is an overprediction of liquid volume, while k_c depresses the predicted volume, Given that the equation overpredicts the liquid volume if both interaction constants are zero, it is evident that a positive non-zero value of k_c will improve liquid volume prediction. The assignation of these interaction constants to small values (0.05) is purely arbitrary; the size was selected by comparison with the optimised values of interaction constant found by other workers for other cubic equations applied to the same data sets. It should be stressed that no optimisation has been carried out to find the 'best' values of k_a and k_c for these data.

As a final indication of the effect of these interaction constants, Figures 9 and 10 show the prediction of bubble pressure and saturated liquid volume for settings of k_a and k_c of 0.05. It may be seen that these small non-zero values give significantly improved predictions, even without rigorous optimisation.

FIGURE 8: R22/R11 isothermal liquid volumes

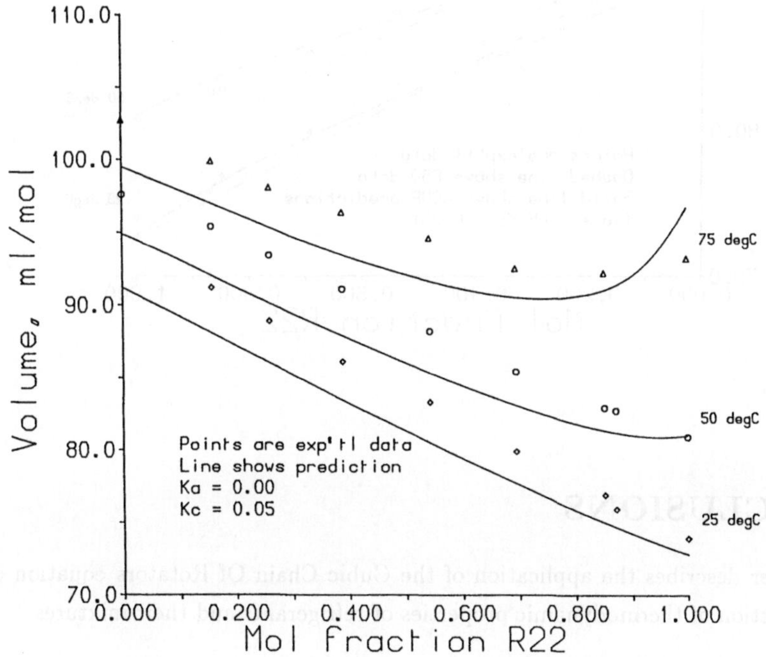

FIGURE 9: R22/R11 isothermal bubble pressures

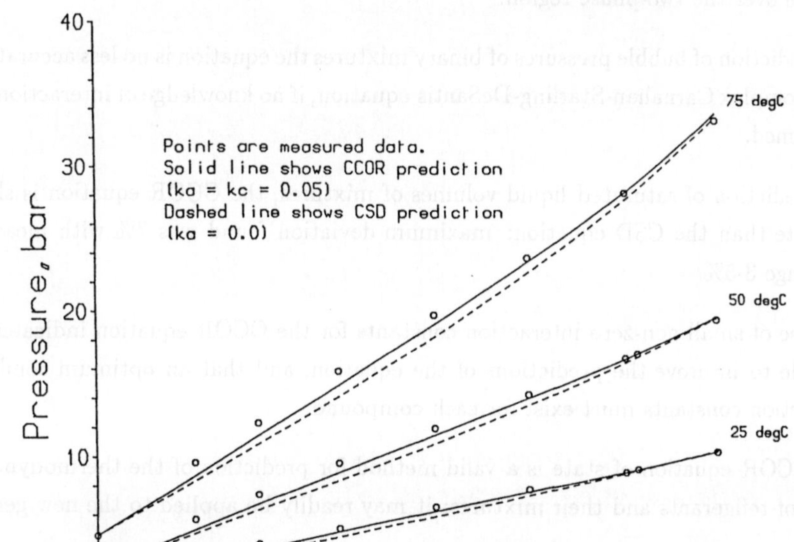

FIGURE 10: R22/R11 isothermal liquid volumes

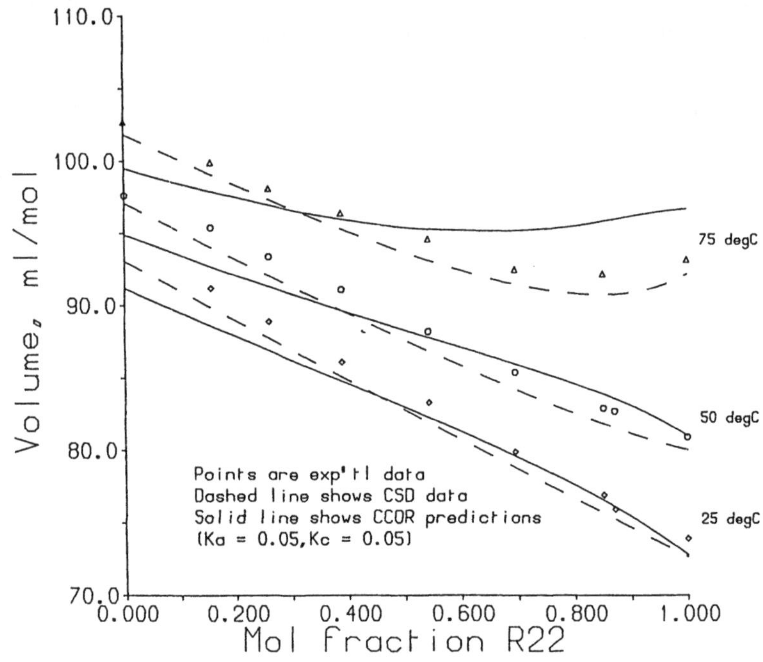

6 CONCLUSIONS

- This paper describes the application of the Cubic Chain Of Rotators equation of state to the prediction of thermodynamic properties of refrigerants and their mixtures.
- The equation predicts pure-component vapour pressures and liquid volumes with accuracies of 1-2% over the two-phase region.
- For prediction of bubble pressures of binary mixtures the equation is no less accurate than the more complex Carnahan-Starling-DeSantis equation, if no knowledge of interaction constants is assumed.
- For prediction of saturated liquid volumes of mixtures, the CCOR equation is slightly less accurate than the CSD equation: maximum deviation found was 7% with mean errors in the range 3-5%.
- The use of small non-zero interaction constants for the CCOR equation indicates that it is possible to improve the predictions of the equation, and that an optimum combination of interaction constants must exist for each compound.
- The CCOR equation of state is a valid method for prediction of the thermodynamic properties of refigerants and their mixtures: it may readily be applied to the new generation of refrigerant fluids.
- The predictions of the CCOR equation of state are slightly more accurate than those found by other workers, using other generalised cubic equations of state

7 NOMENCLATURE

a, b, c, d, c^R	parameters of CCOR equation	
Cp,Cv	heat capacity	kJ/kmolK
f	fugacity	kPa
h,H	enthalpy	kJ/kmol
n	number of mol	mol
N	number of components	
P	pressure	kPa
R	Gas Constant	kJ/kmolK
s,S	entropy	kJ/kmolK
T	temperature	K
U	internal energy	kJ/kmol
v,V	volume	m^3/kmol
x	mol fraction in liquid phase	
y	mol fraction in vapour phase	
z	mol fraction (phase immaterial) *or* compression factor	
Z	compression factor (Pv/RT)	

Greek letters

α	temperature function of CCOR equation
γ	temperature function of CCOR equation
ϕ	fugacity coefficient
Ω_M	parameter of CCOR equation (M one of $a, b, c, d,$)

Sub- and Superscripts

^{o}	perfect gas state
$'$	temperature derivative
c	critical state
i,j,k	component indices
l	liquid phase
p	constant pressure
v	constant volume *or* vapour phase

References

[1] T. Gundersen. *Numerical Aspects of the Implementation of Cubic Equations of State in Flash Calculation Routines.* Comput. & Chem. Eng. 1982 v6 n3 245-255

[2] H. Kim; H-M. Lin; K-C. Chao. *Cubic Chain-Of-Rotators Equation of State* Ind. Eng. Chem. Fundam. 1986 v25 n1 75-84

[3] W. A. Leet; H-M. Lin; K-C. Chao. *Cubic Chain-Of-Rotators Equation of State II for Strongly Polar Substances and their Mixtures.* Ind. Eng. Chem. Fundam. 1986 v25 n4 695-701

[4] H-M. Lin; T. M. Guo; K-C. Chao. *Cubio Chain-Of-Rotators Equation of State and VLE Calculations.* Fluid Phase Equilibria 1983 v13 143-152

[5] R. E. Low. *Techniques for Thermodynamic Property Prediction with reference to Microcomputer Applications.* Proc. 2nd Int. Workshop on Advanced Heat Pumps, Graz 1988 (Tech. Univ, Graz) 329-348

[6] P. M. Mathias; J. F. Boston; S. Watanasiri. *Effective Utilisation of Equations of State for Thermodynamic Properties in Process Simulation.* AIChE Journal v30 182-186

[7] M. McLinden; G. O. Morrison. *A Hard-Sphere Equation of State for Refrigerants and Refrigerant Mixtures* NBS Technical Note 1226 Aug. 1986

[8] M. Meskel-LeSavre; D. Richon; H. Renon. *Bubble Pressures and Saturated Liquid Molar Volumes of Trichlorofluoromethane - chlorodifluoromethane mixtures. Representation of Refrigerant mixtures' Vapor- Liquid Equilibrium data by a modified form of the Peng-Robinson equation of state.* Fluid Phase Equilibria 1982 v8 37-53

[9] G. O. Morrison; M. McLinden. *Two-Refrigerant Mixtures and the Hard-Sphere Fluid.* ASHRAE Trans. 1985 v91 n2 929-943

[10] D. R. Pesuit. *Binary Interaction Constants for Mixtures with a Wide Range in Component Properties.* Ind. Eng. Chem. Fundam. 1978 v17 n4 235-242

[11] R. H. Perry; D. Green (ed). *Perry's Chemical Engineer's Handbook, 6th Edition* McGraw Hill International, New York 1986.

[12] B. E. Poling; E. A. Grens; J. M. Prausnitz. *Thermodynamic Properties From a Cubic Equation of State: Avoiding Trivial Roots and Spurious Derivatives.* Ind. Eng. Chem. Proc. Des. Dev. 1981 v20 n1 127-130

[13] C. L. Pritchard; R. E. Low. *A Self-Regulating Heat Pump for Systems With Variable Power Input.* Proc. 2nd Int. Workshop on Advanced Heat Pumps, Graz 1988 (Tech. Univ. Graz) 265-275

[14] C. L. Pritchard; R. E. Low. *A Self-Regulating Heat Pump for Renewable Energy Applications.* To be presented at the BHRA Int. Conf. on Heat Pump Systems, Munich 1990

[15] M. Shacham. *An Improved Memory Method for the Solution of a Non-Linear Equation.* Chem. Eng. Sci. 1989 v44 n7 1495-1501

[16] Y. Shigaki; H. Orimo et al. *Estimation of the Basic Properties (T_b, T_c, P_c and ω) of Fluorocarbon Refrigerants (Freons).* Int. Chem. Eng. 1988 v28 n3 447-454

[17] R. J. Topliss; D. Dimitrelis; J. M. Prausnitz. *Computational Aspects of a Non-Cubic Equation of State for Phase Equilibrium Calculations: Effect of Density-Dependent Mixing Rules.* Comput. & Chem. Eng. 1988 v12 n5 483-489

APPENDIX 1: THERMODYNAMIC PROPERTIES FROM THE CCOR EQUATION OF STATE

Parameters of the CCOR equation

The CCOR equation has five parameters for each fluid: a, b, c, d, c^R. These may all be obtained from the critical temperature, critical pressure and acentric factor, using the following relations:

$$\begin{aligned}
a &= \alpha(T_r).\Omega_a(RT_c)^2/P_c && (9)\\
b &= \Omega_b RT_c/P_c && (10)\\
c &= \gamma(T_r).\Omega_c RT_c/P_c && (11)\\
d &= \Omega_d(RT_c)^2/P_c && (12)\\
c^R &= 24.863\omega - 33.368\omega^2 + 57.266\omega^3 \ (\omega \geq 0) && (13)\\
&= 0 \ (\omega < 0)
\end{aligned}$$

The Ω factors are defined by the following relations:

$$\Omega_b = 0.4756 - 3.396z_c + 8.236z_c^2 \quad (14)$$

$$\Omega_c = 0.42\Omega_b + 1 - 3z_c \quad (15)$$

$$\Omega_a = \Omega_c(1 + 0.42\Omega_b) + \Omega_b(0.77 + 0.055c^R) + 3z_c^2 \quad (16)$$

$$\Omega_d = \Omega_c(0.77 + 0.055c^R) + 0.42\Omega_a - z_c^3/\Omega_b \quad (17)$$

The temperature dependence of the parameters a and c is described by the functions $\alpha(T_r)$ and $\gamma(T_r)$:

$$\alpha(T_r) = \left(1 + A_1(T_r^{1/4} - 1) - A_2(1 - T_r^2)\right)^2 \ (T_r \leq 1) \quad (18)$$

$$= exp\left(A_3(T_r - 1) - A_4(\sqrt{T_r} - 1)\right) \ (T_r > 1)$$

$$\gamma(T_r) = exp\left(C_1(1 - T_r^{C_2})\right) \quad (19)$$

where:

$$A_1 = 2.66709 + 5.51328\omega - 2.65333\omega^2 \quad (20)$$

$$A_2 = 0.18471 + 0.38357\omega - 0.32706\omega^2 \quad (21)$$

$$A_3 = 0.38349 + 4.88358\omega + 14.2113\omega^2 \quad (22)$$

$$A_4 = 1.17543 + 14.3109\omega + 30.5032\omega^2 \quad (23)$$

$$C_1 = 7.04333 - 5.00422\omega + 1.88597\omega^2 \quad (24)$$

$$C_2 = 0.23177 + 0.16896\omega + 0.03179\omega^2 \quad (25)$$

Equation of State:

$$P = \frac{RT}{v - 0.42b} + \frac{RTb\left(0.77 + 0.055c^R\right)}{v\,(v - 0.42b)} - \frac{a}{v\,(v + c)} - \frac{bd}{v\,(v + c)\,(v - 0.42b)} \quad (26)$$

Fugacity Coefficient:

$$\log\left(f/P\right) = \log\,\phi = Z - 1 - \log\,Z - \int_\infty^V \left[\frac{Z - 1}{V}\right] dV_T \quad (27)$$

which gives, for the CCOR equation:

$$\log\,\phi = \left[\frac{d}{0.42RT(c + 0.42b)} - \frac{1.19 + 0.055c^R}{0.42}\right] \log\left(\frac{v - 0.42b}{v}\right) + \left[\frac{a(c + 0.42b) - bd}{RTc(c + 0.42b)}\right] \log\left(\frac{v}{v + c}\right) + Z - 1 - \log\,Z \quad (28)$$

Internal Energy Departure Function:

$$\Delta U = U - U^\circ = \int_\infty^V \left[T\left(\frac{\partial P}{\partial T}\right)_V - P\right] dV_T \quad (29)$$

The result for the CCOR equation is:

$$\Delta U = \frac{1}{c}\left[a - T\left[a' - \frac{ac'}{c}\right] - \frac{Tbdc'}{c(c + 0.42b)}\left[\frac{1}{c} + \frac{1}{c + 0.42b}\right]\right] \log\left(\frac{v + c}{v}\right) + \frac{d}{0.42(c + 0.42b)}\left[\frac{Tc'}{c + 0.42b} + 1\right] \log\left(\frac{v - 0.42b}{v}\right) + \frac{Tc'}{c(v + c)}\left[a - \frac{bd}{c + 0.42b}\right] \quad (30)$$

The enthalpy and entropy departure functions may be found directly from the internal energy departure functions by the simple relations:

$$\Delta H = \Delta U + Pv - RT \quad (31)$$

$$\Delta S = (\Delta H - RT\log\phi)/T \quad (32)$$

The departure functions may be used to find the enthalpy or entropy of the fluid in any state relative to a given datum state, if the perfect gas heat capacity is known as a function of temperature:

$$H - H^* = \Delta H(v,T) + \int_{T^*}^{T} Cp^\circ\, dT - \Delta H(v^*,T^*) \quad (33)$$

$$S - S^* = \Delta S(v,T) + \int_{T^*}^{T}\left[\frac{Cv^\circ}{T}\right] dT + R\log\left(\frac{v}{v^*}\right) - \Delta S(v^*,T^*) \quad (34)$$

Fugacity coefficient of a Mixture Component

$$\log(f/Pz_i) = \log\phi_i = -\frac{1}{RT}\int_{\infty}^{V}\left[\left(\frac{\partial P}{\partial n_i}\right)_{T,V,n_j} - \frac{RT}{V}\right] dV_T - \log Z \quad (35)$$

The resulting formula derived from the CCOR equation is:

$$\begin{aligned}\log\phi_i = & -\log Z \quad (36)\\ & - \left[2.833 + 0.13095c_{ii}^R + \frac{1}{0.42bRT(c+0.42b)}\left[\frac{d(0.42\Sigma_b + \Sigma_c)}{c+0.42b} - \Sigma_d\right]\right]\\ & \times \log\left(\frac{v-0.42b}{v}\right)\\ & + \frac{1}{RTc}\left[\Sigma_a - \frac{a\Sigma_c}{c} + \frac{1}{c+0.42b}\left[bd\left[1+\Sigma_c\left[\frac{1}{c}+\frac{1}{c+0.42b}\right] + \frac{0.42\Sigma_b}{c+0.42b}\right] - b\Sigma_d - d\Sigma_b\right]\right]\\ & \times \log\left(\frac{v}{v+c}\right)\\ & + \frac{\Sigma_b}{v-0.42b}\left[1.19 + 0.055c^R - \frac{d}{RT(c+0.42b)}\right] + \frac{\Sigma_c}{RTc(v+c)}\left[\frac{bd}{c+0.42b} - a\right]\end{aligned} \quad (37)$$

The constants a, b, c, d, c^R for the mixture are calculated using the van der Waals' mixing rule, which is:

$$M_m = \sum_{i=1}^{N}\sum_{j=1}^{N} z_i z_j M_{ij} \quad (38)$$

The "cross constants" $(M_{ij}, i \neq j)$ are found using the relations below:

$$a_{ij} = (1-k_a)(a_{ii}a_{jj})^{1/2} \quad (39)$$

$$b_{ij} = (b_{ii} + b_{jj})/2 \quad (40)$$

$$c_{ij} = (1-k_c)(c_{ii} + c_{jj})/2 \quad (41)$$

$$d_{ij} = (d_{ii}d_{jj})^{1/2} \quad (42)$$

$$c_{ij}^R = (c_{ii}^R + c_{jj}^R)/2 \quad (43)$$

The notation $\sum_M$ in the relation for component fugacity coefficient is shorthand for the quantity $2\sum_{k=1}^{N} z_k M_{ik}$, evaluated for component i. M may be any one of the five CCOR parameters.

The enthalpy and entropy of a mixture relative to the equivalent enthalpy and entropy of a hypothetical system of the unmixed components at their datum states are given by formulæ analagous to those for the pure component:

$$H_m - \sum_{i=1}^{N} x_i H_i^* = \Delta H_m (v_m, T) + \sum_{i=1}^{N} x_i \left[\int_{T_i^*}^{T} Cp_i^\circ \, dT - \Delta H (v_i^*, T_i^*) \right] \tag{44}$$

$$\begin{aligned} S_m - \sum_{i=1}^{N} x_i S_i^* &= \Delta S_m (v_m, T) + \sum_{i=1}^{N} x_i \left[\int_{T_i^*}^{T} \frac{Cv_i^\circ}{T} \, dT - \Delta S (v_i^*, T_i^*) \right] \\ &+ R \sum_{i=1}^{N} x_i \log \left(\frac{v_m}{v_i^*} \right) - R \sum_{i=1}^{N} x_i \log x_i \end{aligned} \tag{45}$$

Chapter 9

ETHYLCHLORIDE: A VIABLE ALTERNATIVE FOR MEDIUM AND HIGH-TEMPERATURE COMPRESSION HEAT PUMP CYCLES

M Narodoslawsky, J Fresner, K Bärnthaler and F Moser (Graz University of Technology, Austria) J Röhler (Shell PAE Laboratory, Hamburg, Germany)

ABSTRACT

Worries about global warming trends have brought back interest in energy saving techniques and with that in heat pumps. However, with ozone depletion also very much in the minds of decision makers in industry and politics, compression cycles operated with fully halogenated compounds have a serious drawback in the competition for energy saving technologies.
From thorough thermodynamic and toxicologic screening of about 940 substances widely used in chemical engineering, ethylchloride emerged as a potential replacement for R114 in compression cycles. Major advantages of this of this substance include low (and known!) toxicity, compatibility with oils and materials of construction and both, better performance parameters and wider field of application than R114. Given the high contend of hydrogen in the molecule, atmospheric lifetime of ethylchloride is low and hence low or zero ozone depletion and greenhouse effect are expected.
First results in an experimental test rig, corroborating the theoretical findings, will be presented as well as results of material compatibility tests. In these tests, ethylchloride shows high potential for posing a viable alternative for medium and high temperature applications of compression heat pump cycles.

Heat Pumps – Proceedings of the 4th International Conference, pp.101-114

LIST OF SYMBOLS

COP	Coefficient of performance
P	Pressure
P_{shaft}	Shaft power
Q	Heat
q_v	Volumetric heat capacity
T	Temperature
π	Pressure ratio

Subscripts

boil	Boiling point
comp,out	Compressor outlet
cond	Condenser
crit	Critical
eva	Evaporator
sup	Superheating

INTRODUCTION

In recent years the underlying rational for installation of heat recovery systems has changed considerably. In the seventies and early eighties, after two oil shocks, the saving of energy costs was in the center of interest. The oil glut of late has considerably reduced the incentives to save energy under these aspects. However, this trend is counterbalanced by the need to reduce noxious compounds generated by the combustion of fossil energy sources (e.g. the oxides of sulfur and nitrogen). On top of this, reduction of carbon dioxide in order to prevent global warming gets ever more priority and may well become the single most important driving force behind energy saving. This new situation calls for heat pump installations which are ecologically acceptable.

Compression cycle media used in heat pumps are generally of the CFC type. These compounds are known to play a detrimental effect on the high altitude ozon layer protecting the earth from too high exposure to the ultraviolet part of sunlight. R12, which is widely used in heat pumps, is therefore subject to the Montreal Protocol of 1987. A reduction to 50% of 1986 levels of production and use of this substance until 1999 has been mandated for countries who ratified this protocol. Future restrictions in the course of the revision process of the Montreal Protocol might even be more stringent. No restrictions, however, have been imposed on R22, which is the second important compression heat pump medium.

On top of their ozone depleting nature, CFC media are also contributing to the global warming trend. Molecules of R12 and

R22 are calculated to be about 18.000 times, respectively 2.000 times as effective as a carbon dioxide molecule. However, due to the much smaller tonnage of CFC substances released to the atmosphere compared to carbon dioxide, the overall effect on the global warming trend is only about 15%.

However, it should be mentioned that heat pumps contribute only a small portion to global CFC consumption. According to Kuijpers (1989), consumption of restricted CFC substances for heat pumps was approximately 600 t in 1987, with an additional consumption of R22 in the same size.

This new situation makes necessary a new approach to heat pump development, taking into account new requirements for working media and the call for wider applicability of heat pumps, especially in industrial processes. Long dormant interest into new media not belonging to the CFCs as well as research into new cycles extending the range of applicability of well known media like water and ammonia are revived again. This work deals with ethylchloride, a newly found "old" alternative to CFC working media for medium and high temperature industrial heat pump applications.

THE SEARCH FOR NEW COMPRESSION CYCLE MEDIA

A working medium for compression cycle heat pumps has to fulfill stringend requirements. The most important ones are :

- High efficiency in order to save as much primary energy as possible.
- Safe operation and enviromental acceptability.
- Technically suitable with cheap materials of construction.

These different requirements call for different approaches in the search for new working media. The requirement of high efficiencies (also including the call for small and therefore cheap equipment), is best investigated via a theoretical screening on the base of thermodynamic properties.

The second requirement makes the evaluation of the toxicity of all alternatives necessary. This can be done quite easily for components for which the influences on human health are known. In that case a thorough investigation of relevant literature and data banks will come up with answers to most questions. However, if no data are available on health hazards posed by a certain substance, expensive and time consuming tests are necessary. These tests delay introduction on the marketplace for several years. This is surely a drawback for complete newcommers.

Another important aspect is flammability. One of the main arguments for the use of CFC compounds in heat pumps has been their non-flammable character. Albeit this is an important property it should be kept in mind, that today`s technical abilities may ensure safe operation even with flammable substances. There is a marked difference in valuation between toxicity or enviromental acceptability and flammability. Though the latter should be kept low, a substance deemed toxic or detremental to the enviroment does not stand a high chance of introduction on large scale, even if secure operation is technically ensured.

The toughest problem in this respect is clearly the estimation of ecological impacts posed by a substance. The database on this subject is scarce or non existent. A possible approach is, however, to look for a non toxic substance with low atmospheric lifetime. Though this is no guarantee for ecological acceptability, the chance of such a substance to be ecologically harmful is almost nil.

The third group of requirements, calling for compatibility with common materials of construction, can only be investigated by experiment. Therefore, these requirements should be tested only at a stage when the number of candidate-media is low enough to allow for these experiments.

Thermodynamic Search

A search for thermodynamic suitable media for compression cycles is conveniently done by help of chemical engineering "flowsheet programs". This type of programs usually comprise a broad thermodynamic database and the possibility to use various data generators (e.g. different equations of state). Unterleutner (1983) proposed to use critical data as a base for comparision for different substances with respect to their suitability as compression heat pump media. With the help of the flowsheet program PROCESS(R) criteria for optimal working media have been derived. Table 1, which was taken from Narodoslawsky (1988), shows these criteria. They compare favorably with others, proposed by McLinden (1988).

Another result from the evaluation of criteria for optimal media was, that the main difference between substances occures in the volumetric heat capacity. The coefficient of performance (COP), in contrast, is fairly comparable for most substances, given their critical temperature is not too low. Therefore it is appropriate to look for a substance with high volumetric heat capacity showing a reasonably high COP. A sample of approximately 40 out of 940 substances which exhibit optimal performance data was subsequently subjected to closer investigations. Almost all these substances promised better performance parameters than conventional CFC media in medium and high temperature applications, with condensing temperatures between 70 and 150 °C.

Table 1: Thermodynamic criteria for optimal compression cycle heat pump media

CHP performance parameter	Critical property	Optimal value	Influence
COP	T_{crit}	High	Generally weak*)
	P_{crit}	High	Sizeable
	Acentric factor	±	Weak
Volumetric heat storage	T_{crit}	Low	Strong
	P_{crit}	High	Sizeable
	Acentric factor	0,2 or lower	Strong
Pressure ratio	T_{crit}	Low	Strong
	P_{crit}	±	Weak
	Acentric factor	Low	Strong

*)Sharp decrease of COP if reduced condenser temperature is less than 0.96

Toxicity And Ecological Acceptability

The 40 remaining substances were first tested according to their respective toxicity. In this course, all carcinogenic, mutagenic and teratogenic (i.e. hazardous for the human fetus) substances were thrown out. Also all acutly toxic substances with limit threshhold values lower than 100 ppm were deleted from the sample. This left three substances, namely ethylchloride, ethylbromide and propylchloride in consideration.

There exist no data on the impact on the ecological system of these three substances. However, the molecules of all three contain high amounts of hydrogen. This indicates that their respective atmospheric lifetimes are quite low. It can therefore be concluded that they do not pose a hazard to the high altitude ozone layer. On top of this a low impact on global warming is anticipated.

In the case of ethylchloride, data indicate that this substance does not pose a hazard on aquatic life.

Technical Suitability

To examine the technical suitability of a substance to be used in a vapour compression cycle intensive experimental work is necessary.

First tests were made to investigate compatibility of ethyl-

bromide and propylchloride with oils and copper as well as steel. These tests followed the ASHRAE P97 recommandations and were carried out at the Shell PAE labors in Hamburg/FRG. These tests were carried out at a temperature of 175 °C and lasted 14 days. Shell oil CLAVUS G 100 and BITZER oil 100 were used in the tests. Ethylbromide lacked stability under these conditions and attacked the steel and copper samples. Both oils were visibly darkened and fall out was observed. Thus, ethylbromide is no alternative for CFC working media for heat pumps.
Propylchloride did not attack the metals and both oils proved to be applicable, with slight advantages for the Shell product. The chloride and the oil did not form sparate liquid phases. For technical reasons, no tests were made with ethylchloride. However, the chemical similarity between the two chlorides indicated, that ethylchloride may also be a viable alternative. This could be corroborated by results of further tests that will be reported in the course of this work.
Since ethylchloride is applicable in the more interesting temperature range with condensing temperatures between 70 and 120 °C further investigations were centered on this substance.

ETHYLCHLORIDE AS A HEAT PUMP MEDIUM

Ethylchloride is a rather wellknown substance. Its existence has been reported as long ago as 1500 A.D. and it was available in pure form around 1830. Henning (1920) reports that before World War I production in Germany exceeded 350 metric tons a year. Its main application was as ethylating agent in the production of colours.
Besides that it was used as an anaesthetic, both for local use and for inhalation.
Another not quite surprising use for ethylchloride was as working medium in refrigeration units. Until the advent of CFC media it was considered a safe and efficient working medium. Namely in the USA, ethylchloride was used in various installations, including mobile refrigeration units and refrigeration units on board of warships. Its low pressure made it preferable to ammonia in these cases.

Thermodynamic Data

There are not very many data on the thermodynamic behavior of ethylchloride available in literature. The two most important papers on this subject were published by Gallant (1966) and Das (1968). Figure 1 gives a temperature vs. entropy diagram which was established using the Benedict-Webb-Rubin equation of state

(1940) in the extension of Starling and Han (1972). It agrees well with data from the afore mentioned authors. The critical data and the normal boiling point are given in table 2.

Table 2: Thermodynamic data of ethylchloride

T_{crit}	[°C]	186.2
P_{crit}	[bar]	52.7
acentric factor		0.19
T_{boil}	[°C]	12,4

Toxicity and Flammability

As mentioned above, ethylchloride is neither carcinogenic nor mutagenic nor teratogenic. This greatly reduces any longtime health hazards and increases its ecological acceptability.
Besides this, ethylchloride has low acute toxicity. Its limit threshold value according to Kühn-Birret (1984) is 1000 ppm. This is the highest value bestowed on any compound that is not a natural constituent of the atmosphere.
Ethylchloride is inflammable. Jones (1928) reported the concentration range over which ethylchloride and air form an inflammable mixture to be 4 to 14.8 % (volume). In this respect it is also noteworthy, that because of its high latent heat ethylchloride does not evaporate fast, thus reducing the hazard of fire by bad handling or leakage.

Compatibility With Materials of Construction

The first tests at the Shell PAE labors indicated, that copper and steel may be safly used in an installation employing ethylchloride. First runs in a test rig which was constructed from these materials reiterated these findings. However, after a revamping of the test plant and the installation of a new evaporator which also contained aluminia, severe corrosion was observed. Therefore new compatibility tests had to be made to get more information on possible material problems. Table 3 presents the test conditions for these experiments. All samples were held at a uniform temperature of 125 °C in sealed steel tubes. The colour of the ethylchloride/oil mixture and the appearance of the metals was inspected by vision once the sealed tubes were unloaded.
The tests corroborated the hypothesis that copper and steel are

not attacked by ethylchloride, and that the oil/refrigerant mixture is stable.

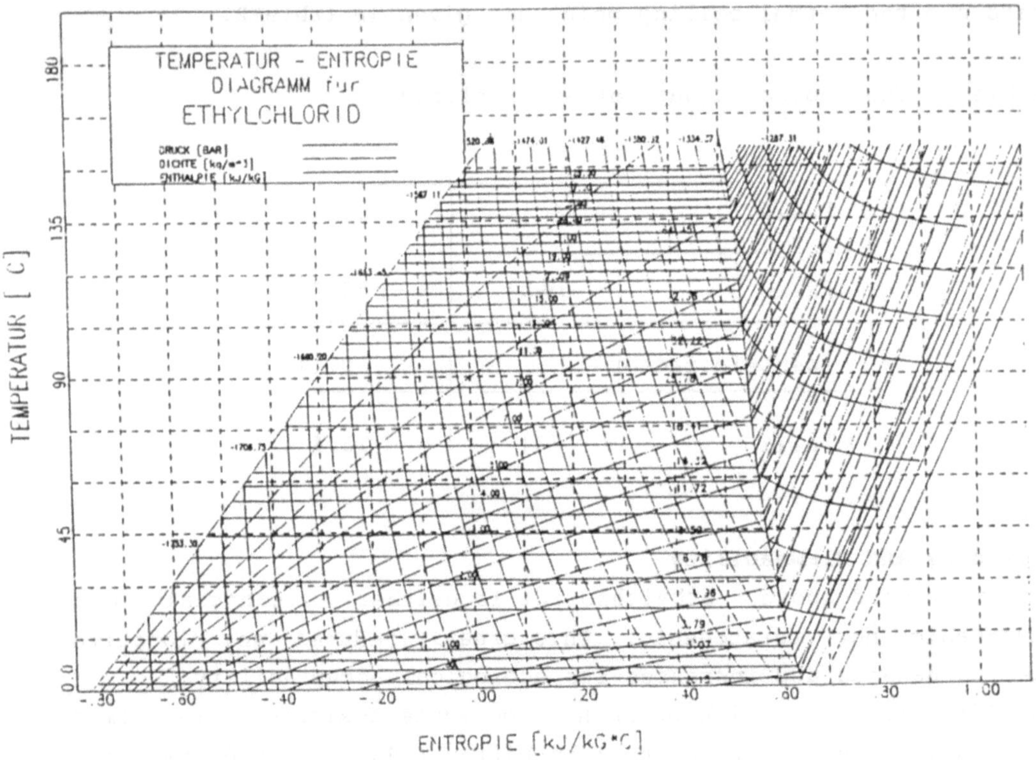

Figure 1: Temperature vs. entropy diagram for ethylchloride

Table 3: Test conditions for material tests

	TEST 1	TEST 2	TEST3	TEST4	TEST5	TEST6
IRON [g]	9.4	10.3	10.2		4.0	
IRON OXIDE [g]						0.25
COPPER [g]	5.5	5.7	5.7			
ALUMINIUM [g]	1.5	1.6		1.3	1.7	0.51
WATER [g]	2.7			0.5	0.5	
ETHYL CHLORIDE[g]	7.	9.	9.0	1.7	2.5	9
OIL CLAVUSG100 [g]	10.7	10.	9.8	5.0	6.3	3.6
TEMPERATURE [°C]	125	125	125	125	125	125
DURATION OF EXPERIMENT	14 days	14 days	14 days	7 days	7 days	7 days

However, the samples containing aluminium showed severe corrosion. Thus aluminium is not a viable material for use in any installation operated with ethylchloride.
Further tests were carried out to test compatibility with standard insulating lacquer as used in hermetic compressors. No detrimental influences of ethylchloride were observed.
There are of course other materials of construction to be taken into consideration. Most important are desiccants and seals. Both caused some trouble in the test plant. However, given the appli-

cation of ethylchloride as working medium in refrigeration units decades ago, there should be a solution to these problems. Test of possible desiccants without binders (which are dissolved by ethylchloride) look promising but are not yet conclusive.

COMPARISON WITH CFC REFRIGERANTS

An alternative working medium for compression heat pumps has not only to be more ecologically acceptable but also better than refrigerants now in use. Therefore, ethylchloride was extensively compared with other media by simulation as well as in experiment. In a first step several CFC media were considered to perform temperature boosts from 20 to 70 °C and from 60 to 120 °C. These applications were chosen because of their importance in industrial heat recovery tasks. Tables 4 and 5 present the results of this simulation.

Table 4: Comparison between ethylchloride and conventional media for a temperature boost from 20 to 70 °C

Substance	COP	π	q_v (kJ m^{-3})	P_{cond} (bar)	$T_{comp.out}$ (°C)
Ethyl chloride	4.99	4.39	1339	5.9	97
R21	4.97	4.35	1511	6.6	101
R11	4.96	4.54	880	4.0	86
R12	4.39	3.32	3705	18.5	83
R22	4.23	3.33	6025	30.3	98
R114	4.49	4.02	1403	7.3	72
Pentane	4.73	5.06	563	2.8	72
Butane	4.56	3.38	1635	8.0	73

Table 5: Comparison between ethylchloride and conventional media for a temperature boost from 60 to 120 °C

Substance	COP	π	q_v (kJ m^{-3})	P_{cond} (bar)	$T_{comp.out}$ (°C)
Ethyl chloride	4.39	3.82	3467	17.4	138
R114	3.37	3.59	2830	20.5	122
R114B2	4.27	4.36	1182	6.5	122
R21	4.36	3.79	3891	19.5	144

For the lower temperature case ethylchloride clearly comes out first in terms of COP. The only comparable CFC compound is R21 which is, however, reported to be carcinogenic. Though R11 is close with respect to COP it lags far behind in terms of the volumetric heat capacity. All other contenders either exhibit low COP or too high pressures in the condenser, or both.

In the high temperature case only R114 and R21 may be applied. Again, ethylchloride is shown to be superior to these substances. As medium to high temperature heat recovery tasks are the most interesting field of application for ethylchloride, a more thorough comparision to R114 is necessary. Figures 2, 3 and 4 are comparisions of results from simulation calculations of ethylchloride and R114.

Figure 2 shows clearly that the COP of ethylchloride is superior to that of R114. The difference is increasing with evaporator temperature and temperature boost.

The difference in the volumetric heat capacity between the two substances can be seen from fig.3. Again, ethylchloride is superior, and the difference is increasing with evaporator temperature.

Finally, fig.4 shows the comparison of the pressure ratios. Though ethylchloride exhibits slightly higher values it should be kept in mind that its pressure levels are considerably lower for the same evaporator temperatures.

To verify these simulated comparisons, experimental runs were carried out with both refrigerants. A schematic drawing of the test plant is given in fig. 5. More details on the experiments may be found in Bärnthaler (1989). Tables 6 and 7 summarize the results of these experiments. The coefficient of performance was calculated by

$$COP = \frac{Q_{cond}}{P_{shaft}}$$

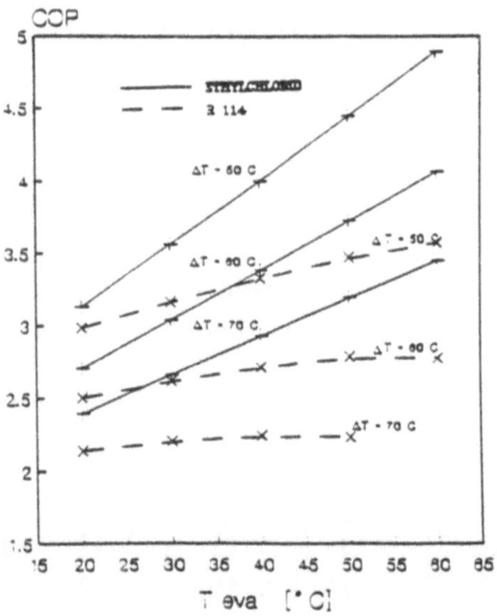

Figure 2: COP dependant on evaporator temperature and temperature boost

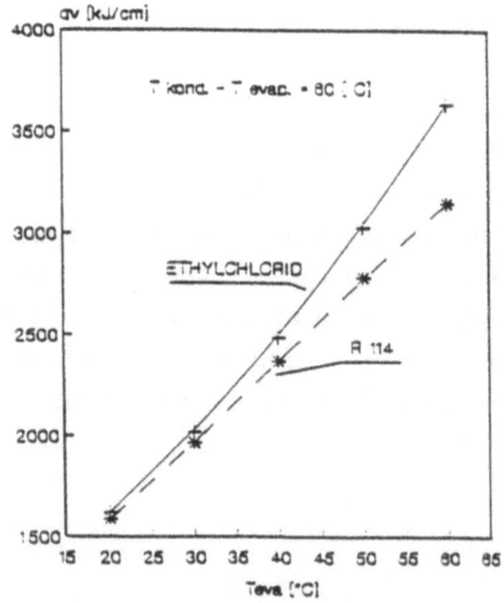

Figure 3: Volumetric heat capacity dependant on evaporator temperature for constant temperature boost (60 °C)

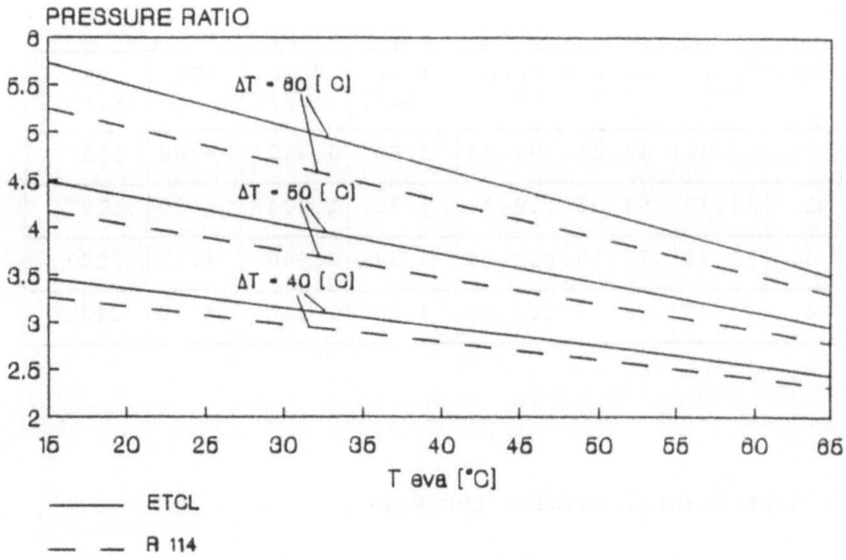

Figure 4: Pressure ratio dependant on evaporator temperature and temperature boost

Carnot`s COP is defined as

$$COP_{carnot} = \frac{T_{cond}}{T_{cond}-T_{evap}}$$

The performance factor n is calculated from

$$n = \frac{COP}{COP_{car}}$$

As can be seen from the tables, the simulated values for the differences in the COP are reinforced by the experiments. The volumetric heat capacity, however, seems to be closer together for both substances. For lower condensing temperatures, a slight advantage for R114 is visible, which vanishes at higher temperatures. This (as well as the somewhat lower than expected differences in the COP) may be attributed to larger heat losses in the compressor when operated with ethylchloride, since this substance exhibits higher compressor discharge temperatures. However, the experiments confirm the trends stated by the simulation calculations.

Table 6: Experimental results for ethylchloride

Exp.	T_{eva} [°C]	T_{con} [°C]	T_{sup} [°C]	p_{con} [bar]	ΣQ_l [kW]	COP	qv [kJ/m³]
1	12.50	44.35	90.93	2.98	0.410	4.08	1157
2	30.70	69.08	109.90	5.90	1.004	4.30	1797
3	33.75	73.00	112.50	6.64	0.940	4.17	2123
4	43.97	38.34	123.65	9.34	1.310	4.03	2432

Table 7: Experimental results for R114

Exp.	T_{eva} [°C]	T_{con} [°C]	T_{sup} [°C]	p_{con} [bar]	ΣQ_l [kW]	COP	qv [kJ/m³]
1	8.73	44.90	71.68	3.66	0.257	3.31	1171
2	31.66	70.70	84.34	7.28	0.497	4.07	1997
3	34.32	75.30	89.26	3.15	0.671	3.79	2076
4	43.48	89.55	104.21	11.25	0.360	3.50	2407

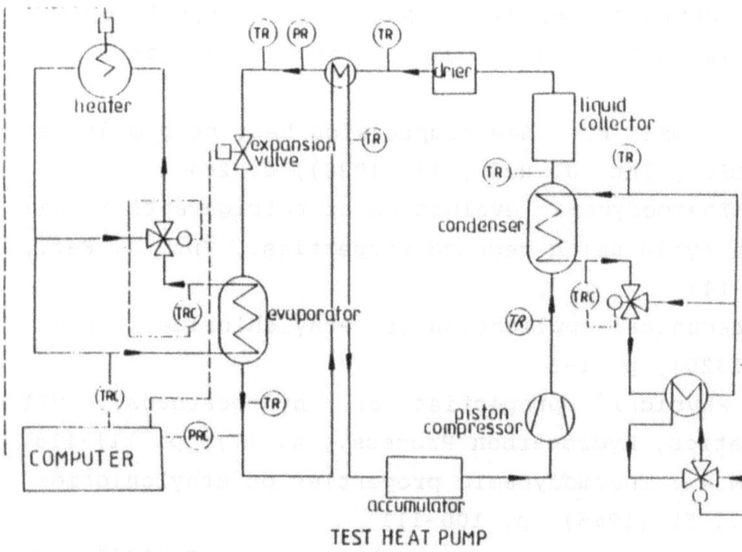

Figure 5: Test plant for media comparison

CONCLUSION

A thorough search for new and ecologically acceptable compression heat pump working fluids has rendered ethylchloride as a viable alternative for medium and high temperature industrial heat recovery tasks. Most tests so far have put an optimistic view on the possibility of introducing this substance as a heat pump medium in the market place in the near future, albeit there are some important features which remain to be investigated. If ethylchloride is really found to be applicable as heat pump medium it might be a contribution to put compression heat pumps back on an environmentally sound road.

ACKNOWLEDGEMENTS

The authors want to express their gratitude to the Austrian Research Council for providing the financial background for this work under the framework of the "Wärmepumpenschwerpunkt S3101". They want also to thank Simulation Sciencies for kindly allowing the use of the simulation program PROCESS(R), which proved to be extremly helpful. Shell PAE labors of Hamburg are also thanked for their very important contribution to this work.

LITERATURE CITED

KUIJPERS L.: Technical progress on protecting the ozone layer; Refrigeration, airconditioning and heat pumps; Technical options report; UNEP, 1989

UNTERLEUTNER K., HUEMER H., MOSER F.: Search for optimal working fluids in high temperature heat pumps; Proc. XVI Congr. of Refrig., Paris 1983

NARODOSLAWSKY M., MOSER F.: New compression heat pump media as replacements for CFCs, Int. J. Ref., 11 (1988), p. 264

MCLINDEN M. O.: Thermodynamic evaluation of refrigerants in the vapour compression cycle using reduced properties., Int. J. Ref., 11 (1988), p. 134-143

HENNING A.: The technical application of ethylchloride, Trans. Farad. Soc., 39 (1920), p. 1-8

GALLANT R.W.: Physical properties of hydrocarbons. VII Chlorinated aliphatics, Hydrocarbon Process., 45 (7), p. 111-118

DAS T.R., KULOOR N.R.: Thermodynamic properties of ethylchloride, J. Ind. Engn. Sci., 50 (1966), p. 100-111

BENEDICT M., WEBB G.B., RUBIN L.C.: J. Chem. Phys., 8 (1940), p. 334

STARLING K.E., HAN M.S.: Hydrocarb. Process. Petrol Refin., 51 (1972), p. 129

KÜHN - BIRRET: Merkblätter Gefährliche Arbeitsstoffe, 6. Aufl., Ecomed, Landsberg/Lech, 1984

JONES G.W.: The flammability of refrigerants, Ind. Eng. Chem., 20 (1928), p. 367-370

BÄRNTHALER K., FRESNER J.: First results of cycle experiments with ethylchloride in compression heat pumps, HP Newsletter, 7 (1989), p. 42-44

PART III

Applications for heat pumps

Chapter 10

POTENTIAL FOR MVR IN INDIAN DISTILLERIES

S Devotta and S Gopichand (National Chemical Laboratory, India), A V Heaton and T Pears (Electricity Council Research Centre, Chester, UK)

ABSTRACT

There is a considerable scope for mechanical vapour compression (MVR) open cycle heat pumps for energy conservation in ethanol distillation in India. In the proposed MVR heat pump assisted distillation scheme, a freshly treated water stream is preheated, expanded and fed to a falling film evaporator. Low pressure steam is produced by extracting heat from the overhead vapour from the rectifier. The low pressure steam thus generated is compressed and then fed to the analyser column along with the make-up steam. The temperature driving force in the falling film evaporator is optimised to its capital cost. The feed is preheated to the required temperature entirely by the sensible heat of the spent wash.

The potential of MVR heat pumps in alcohol distillation with respect to Indian operating and economic parameters is assessed. The necessary modifications required to integrate an MVR to a conventional Indian distillery are described. The other implications with respect to cogeneration, availability of hardware, government fiscal policies and regulations are also discussed.

NOMENCLATURE

A_{UT}	annual net saving on utilities	Rs
C_{UFC}	unit fixed capital cost	Rs $(kl\ d^{-1})^{-1}$
c_E	electricity cost	Rs $(kWh)^{-1}$
c_S	steam cost	Rs t^{-1}
c_P	heat capacity at constant pressure	kJ kg^{-1} K^{-1}
c_V	heat capacity at constant volume	kJ kg^{-1} K^{-1}
k	capacity of the plant	kl d^{-1}
P_1	compressor suction pressure	bar
P_2	compressor discharge pressure	bar
(PBP)	payback period	yr

Heat Pumps – Proceedings of the 4th International Conference, pp.117-126

RR	reflux ratio	dimensionless
S	steam compressed	kg s^{-1}
T_1	saturation temperature at P_1	K or ^oC
T_2	saturation temperature at P_2	K or ^oC
T_{EV}	temperature driving force in evaporator	K or ^oC
w_I	isentropic work of compression	kW
w_A	actual work of compression	kW
x_F	feed concentration	volume %
x_D	product concentration	volume %
x_B	bottoms concentration	volume %
$_I$	isentropic efficiency	dimensionless
$_M$	mechanical efficiency	dimensionless

INTRODUCTION

There are more than one hundred and fifty distilleries in India producing mainly rectified spirit from sugar molasses. In general Indian distilleries fall into two categories, attached and detached to sugar mills. These distillery complexes may also include chemical works. In the attached distilleries, bagasse along with coal or oil is used as fuel in boilers to generate steam. Approximately 2 kg of steam is produced for every kg of bagasse used. The attached distilleries, in general, have cogeneration facilities with back pressure turbines. The inlet pressure is about 22 bar and back pressure is 1 bar. The exhaust steam, in the required amount, is used for alcohol distillation. Sugarcane is a seasonal crop, therefore, the mills operate for about six months in a year. The heat to power ratio varies depending on season or off-season time.

In a detached distillery, steam is usually generated either from coal or oil. In the case of coal based plants, the cost of steam produced depends strongly on the proximity of the distillery to the coal mines. Some detached distilleries also have cogeneration systems depending on their size and the power demand of the complex. Invariably, most of the distilleries still draw power from the grid.

CONVENTIONAL DISTILLATION

A simplified conventional scheme followed in Indian distilleries for the distillation of fermented broth containing 7 to 9 per cent alcohol by volume to produce rectified spirit (95 per cent by volume) is shown in Fig. 1. The fermented wash is preheated in the beer heater from 30° to 65^oC by extracting heat from the overhead product of the rectification column and then further heated to about 85^oC by extracting heat from the spent wash. It has been suggested that this temperature can be raised to about 90^oC or even slightly higher (Paul and Mitra, 1964), but there is a reluctance to implement this suggestion as it could cause more degassification problems and more loss of alcohol with it as the feed is closer to its bubble point. The preheated wash is then fed to the analyser (also referred as beer or wash) column. Live steam is directly sparged to the bottom of the beer column to strip alcohol and aldehyde from the descending wash liquid. The alcohol concentration in the bottoms (spent wash or stillage) is generally around 0.03 per cent by volume.

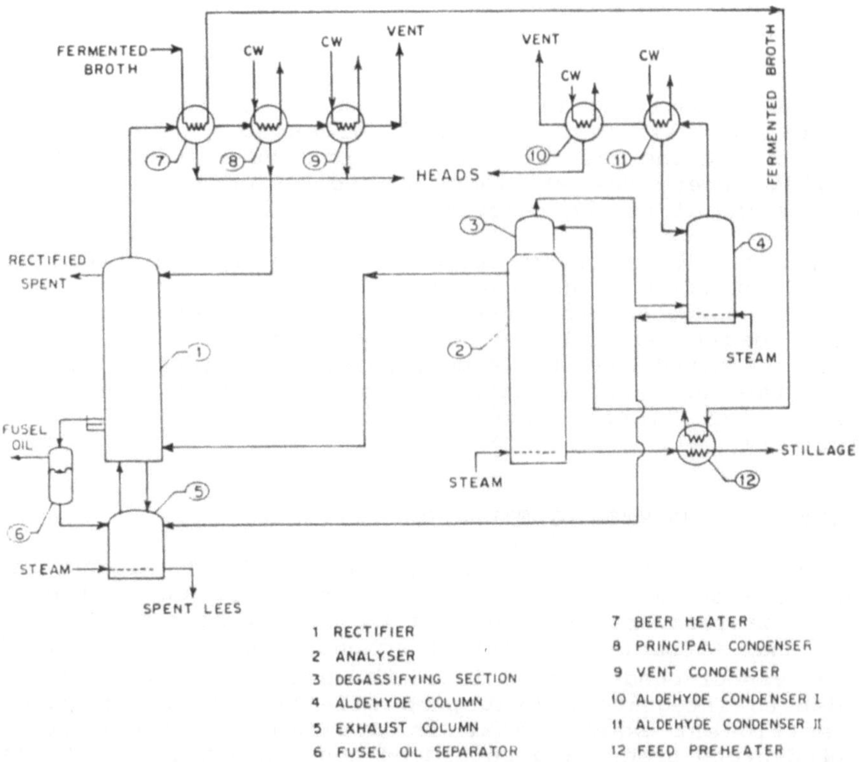

FIGURE 1 : CONVENTIONAL ETHANOL DISTILLATION

The alcohol vapour fraction from the beer column is fed to the rectification column. The vapour from the top of the rectification column is condensed in the beer heater followed by a principle condenser and a vent condenser. The condensates from the condensers are either totally fed back to the top of the column as reflux or some fraction removed as impure alcohol or as heads. Depending on the tray efficiency and the number of plates, the reflux ratio may vary between 3.5 to 4.5.

The product or rectified alcohol is withdrawn 2 to 3 plates below the column head. A mixture of higher boiling components, known as fusel oil, is removed as a bleed from about 4 to 10 plates from the bottom of the rectifier. The downflow from the rectifier is fed to the top of the beer column. More details on the alcohol distillation and some aspects of energy conservation can be found in the references listed at the end of this paper.

Specific Energy Consumption

The factors which determine the specific energy consumption i.e. kg of steam per litre of alcohol are a) reflux ratio, b) heat recovery from spent wash, c) heat recovery from overhead vapour from rectifier and d) heat loss from the plant. With the conventional scheme described above, which includes heat recovery from overhead vapour and spent wash for wash liquor preheating, the specific steam consumption varies between 2.5 and 3.5 kg of steam per litre of alcohol produced. About 40 per cent of heat supplied to the plant is lost through spent wash. Therefore, it is essential that as much heat is recovered as possible from spent wash which is discharged at 80°C.

In some cases, the sensible heat of the spent wash is further utilised to preheat the boiler feed water stream. This is particularly advantageous in independent or detached distilleries where generally the entire steam is consumed in the process and no condensate return is possible. In the case of attached distilleries, such a scheme should be advantageous during off-season. In such cases, the effluent temperature can be brought down to 60°C resulting in an energy saving equivalent to 0.5 kg of steam per litre of alcohol produced. This is also advantageous in the case of plants which have anaerobic treatment of stillage where the stillage has to be cooled down to 37°C before feeding it to the reactor.

The exact specific steam consumption in a distillery for a specified condition is not readily available. Therefore, NCL has developed a computer programme to evaluate the steam consumption from the readily available distillery plant data. From this it has been estimated that for some typical conditions, presented in Table 1, the theoretical steam consumptions are 2.15 and 2.58 kg per litre of alcohol for reflux ratios of 4.0 and 5.0 respectively.

MVR HEAT PUMP ASSISTED DISTILLATION

Nearly half the heat content of the overhead vapour in the conventional scheme is removed by cooling water and wasted. This heat can be recovered by integrating an MVR heat pump. In the conventional alcohol distillation scheme, the major heat load for overhead vapour condensation is in the beer heater and condenser. In order to integrate a heat pump effectively, it is essential that sufficient source heat is available. Therefore, the fermented broth preheating should be done only with spent wash so that the entire heat available with the overhead vapour could be recycled by the heat pump. Conventionally the fermented broth is fed to the analyser at about 85°C. This can conveniently be achieved utilising the sensible heat available in the stillage. This would only require some additional heat transfer area.

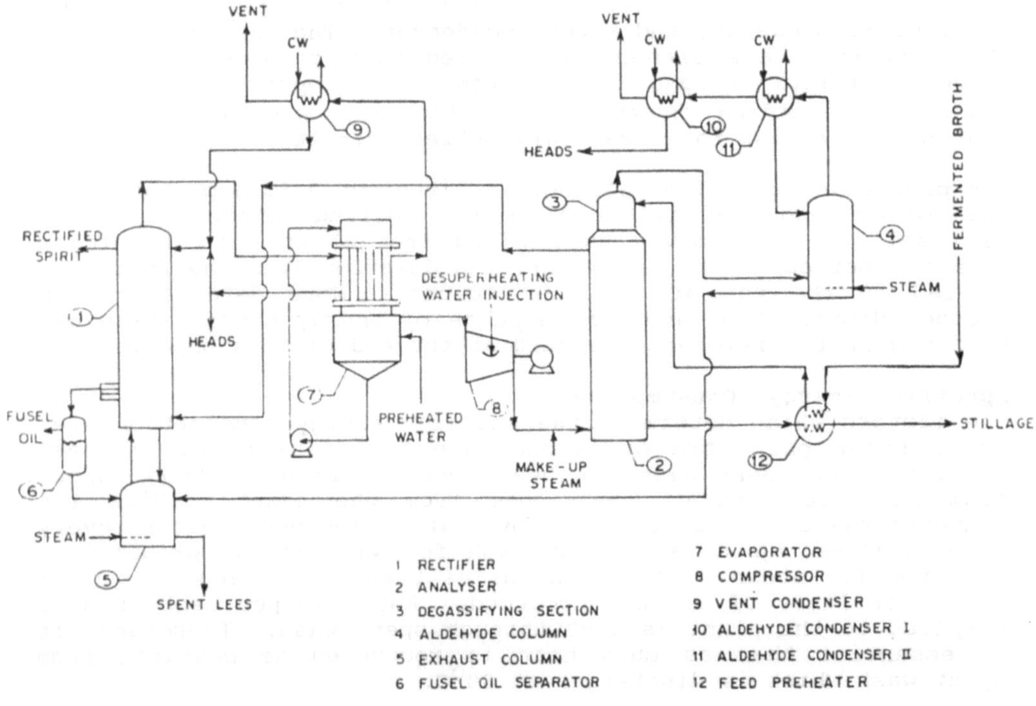

FIGURE 2: MVR ASSISTED ETHANOL DISTILLATION

In an MVR heat pump assisted distillation, either the column overhead or bottoms is used as the working fluid in an open cycle heat pump. In the overhead MVR scheme, the overhead vapour from the rectifier is directly compressed and fed to the heat pump condenser where it delivers heat to generate steam at the required temperature. This steam along with make-up steam is fed to the analyser. The condensed liquid is expanded and then fed to a flash tank. The vapour from the flash tank is fed to the compressor suction port while the liquid is fed to the rectifier as reflux and some part removed as heads.

In the other mode of MVR, known as reboiler flashing MVR, conventionally the bottom product is expanded and then vaporised by extracting heat from the overhead vapour. In order to avoid scaling problems associated with stillage, a freshly treated water stream is preheated with live steam, expanded to a low temperature and pressure and the resultant two phase mixture is vaporised by extracting heat from the overhead vapour from the rectifier as illustrated in Fig. 2. The steam thus generated is then compressed and fed to the analyser column along with make-up steam. The stillage can also be flashed and the steam produced is mixed with the steam generated in the heat pump evaporator and then compressed. However, this arrangement is not considered here as there is no significant advantage over the proposed scheme.

Both MVR schemes eliminate the beer heater and principle condenser but require an evaporator for steam generation. The steam compression can be achieved either by a two stage centrifugal compressor or by a single stage screw compressor. Some of these compressors have been already applied successfully for steam compression. Generally, compressor suppliers are more confident about steam compression than about alcohol compression. Therefore, only reboiler flashing is considered for further discussion.

The live steam to the columns is generally supplied at a pressure of about 2.0 bar. This is perhaps arbitrarily fixed. The column pressure drop is of the order of 0.3 bar while the pressure drop in the sparger is about 0.7 bar. In the case of MVR it is essential that the temperature difference across the column is minimised. Therefore, it is proposed to dispense with the sparger and feed the steam directly over the liquid pool. Although this arrangement leads to the loss of one equilibrium stage, it is not likely to affect the performance of the column.

In the case of MVR, it is essential that the temperature boost across the compressor and the temperature driving force in the evaporator are optimised with respect to the capital cost and the running cost of the MVR heat pump. As the T_{EV} is decreased, capital cost on the evaporator increases, but the vapour density of steam at the compressor suction increases thereby reducing compressor size; the temperature boost across the compressor is also reduced leading to a reduction in the recurring power cost. Usually falling film evaporators specially designed with a low temperature driving force, as low as $3^{o}C$ are used.

TECHNO-ECONOMIC FEASIBILITY

The economic assessment of MVR distillation was made on the basis of simple payback period. The effects of plant capacity and steam, electricity and unit fixed capital costs on payback period were studied. The cost correlations for dry compressors and falling film evaporators were not readily available, but an approximate estimate (in Pounds Sterling, One Pound Sterling is approximately 26 Indian Rupees @ December 1989) of an MVR system to match the capacity of 40 kilo litre per day (denoted by kl d^{-1}) was available. In this estimation, it was assumed that the entire unit would be made in the U.K. However, if expensive components such as, compressor and evaporator are made either fully or partly in India, the cost can be significantly reduced. Also, subsequent units are likely to be still cheaper going down the learning curve. The cost of an MVR system for a 40 kl d^{-1} plant was taken as Rs 6.0 million and the unit fixed capital cost of a plant for any other capacity was estimated using the equation

$$\frac{(C_{UFC})_{k=40}}{C_{UFC}} = (40/k)^{0.66}$$

The power required for isentropic compression was calculated by the equation

$$w_I = S\ c_V\ T_1\ \{(P_2/P_1)^Z - 1\}$$

where

$$Z = (c_P - c_V)/c_P$$

and the actual work required for compression by

$$w_A = \frac{w_I}{(\ I\ \ M)}$$

The simple payback period is calculated by

$$(PBP) = \frac{k\ \ C_{UFC}}{A_{UT}}$$

where A_{UT} is the annual savings on utilities i.e. net difference between the cost of steam saved and the cost of electricity consumed.

Results and Discussion

For a distillery of 40 kl d^{-1} capacity, operating under typical conditions, the integration of MVR can result in a steam saving of the order of 2.65 tonnes per hour at an expenditure of 300 kW of electricity. The scheme is attractive with a payback period of 2.9 years if the steam, electricity and unit fixed capital costs are 250 Rs t^{-1}, 1.25 Rs $(kWh)^{-1}$ and 150,000 Rs $(kl\ d^{-1})^{-1}$ respectively.

It should be noted that the payback periods presented here were calculated without accounting for interest on capital. If this is taken into account at the current rate of 11 per cent (soft loan interest rate applicable in India for energy conservation and modernisation) the payback period is 5.7 years. Additional advantage of lower payback period could be derived if the benefit of 100 per cent depreciation on the initial investment, as per the current Government of India fiscal policies, is exploited. With the above benefit, the payback period falls to a very attractive value of 1.45 years.

The payback period is very sensitive to the cost of steam as can be seen in Fig. 3. It varies from 9.4 yr to 1.98 yr as the steam cost increases from 175 to 300 Rs t^{-1}. It is obvious that the steam cost has to be more than 225 Rs t^{-1}, if a desired payback period of 3.0 yr is to be obtained for a unit fixed capital cost of 150,000 Rs $(kl\ d^{-1})^{-1}$.

The effect of cost of electricity on payback period is shown in Fig. 4. It can be seen that the payback period is very sensitive when the cost of steam is lower than 200 Rs t^{-1}. However, (PBP) is relatively less sensitive when the cost of steam is higher than 300 Rs t^{-1}.

The effect of plant capacity on (PBP) is shown in Fig. 5. The payback period decreases sharply up to the capacity of 50 kl d^{-1} and beyond this capacity the decrease in payback period is only marginal. It may be noted that the most common size of the Indian distilleries is about 30 kl d^{-1}. The payback period is significantly influenced by the unit fixed capital cost and the cost of steam.

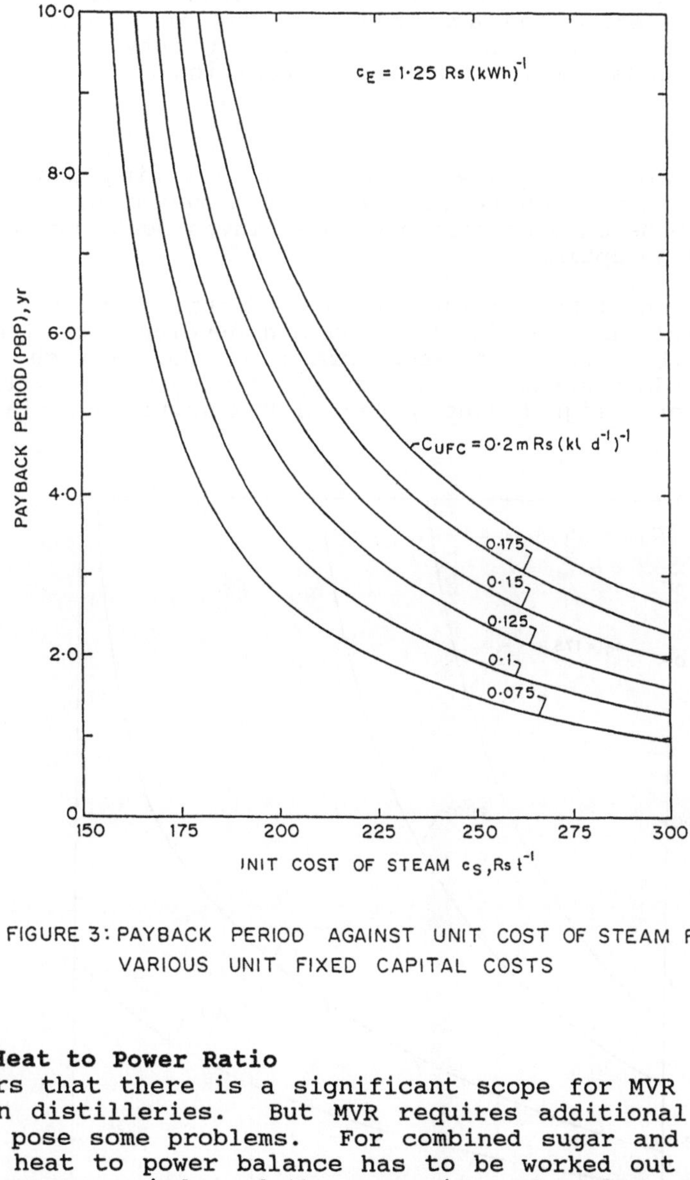

FIGURE 3: PAYBACK PERIOD AGAINST UNIT COST OF STEAM FOR VARIOUS UNIT FIXED CAPITAL COSTS

MVR and Heat to Power Ratio

It appears that there is a significant scope for MVR heat pumps in Indian distilleries. But MVR requires additional power and this may pose some problems. For combined sugar and distillery plants, heat to power balance has to be worked out for season and off-season periods and the economics calculated accordingly. The possibility of steam turbine driven compressors could be considered when there is a restriction posed by the heat to power ratio of the plant. This also keeps the plant free from grid power failures. But, the addition of turbines increases the capital cost. There is no universal answer to this imbalance as this is dictated by the current facilities and practices in the plant under study. The economics may vary from plant to plant mostly influenced by the cost of utilities and the size of the plant. Therefore, it is better that each plant is given an independent study.

Potential for MVR

There are many factors which pose hurdles to the immediate absorption of MVR technology into Indian industry. These include nonavailability of suitable compressors, high import duty on hardware (currently it is about 85 per cent), inadequate fiscal benefits (currently expenditure on energy conservation qualifies for 100 per cent depreciation in the year of investment) and lack of demonstration schemes or venture capital fund (such as the ones available in EEC countries and U.S.A.). All these problems are being looked into by various modes and by suitable interaction with the agencies concerned. Currently some Indian industries are on the verge of manufacturing steam compressors which will expedite the acceptance of MVR in India. If so, there will be significant saving on import duty thereby making MVR a more economically attractive proposition.

CONCLUSIONS

Though MVR offers adequate economics in existing distilleries, the best economics can be achieved for a new plant or a plant going for expansion when size and total investment for the entire system could be optimised.

A demonstration of MVR assisted ethanol distillation in an Indian distillery will be jointly designed and developed by NCL, India and ECRC, U.K. under the sponsorship of the British Overseas Development Administration. It is expected that this will be the forerunner for a significant number of MVR units in India.

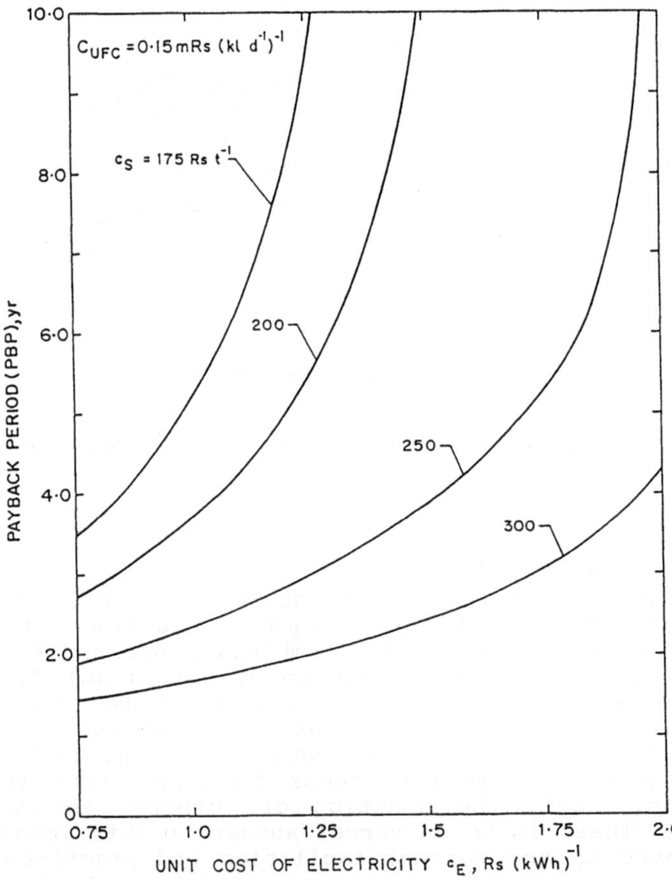

FIGURE 4: PAYBACK PERIOD AGAINST UNIT COST OF ELECTRICITY FOR VARIOUS UNIT COSTS OF STEAM

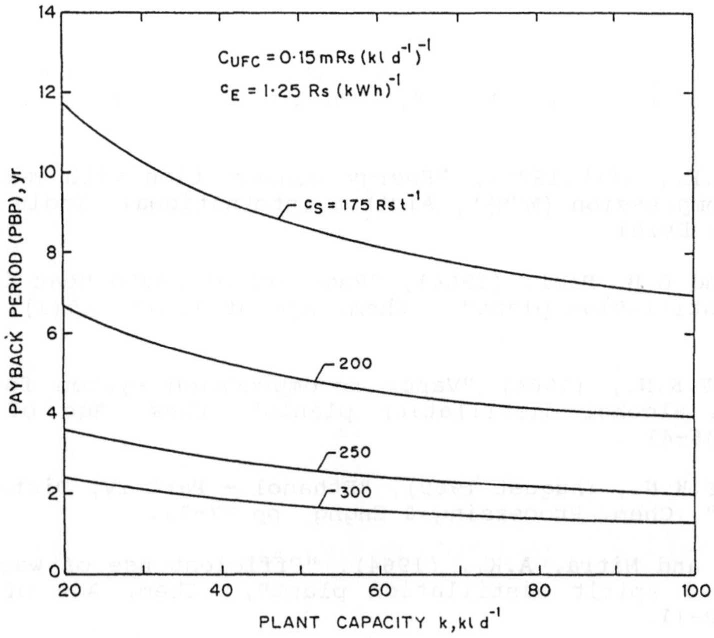

FIGURE 5: PAYBACK PERIOD AGAINST PLANT CAPACITY FOR VARIOUS UNIT COSTS OF STEAM

variables	typical value (range considered)
Process Variables :	
capacity of the plant	40 kl d^{-1} (20 - 100)
bottoms concentration	0.3 % by volume
feed concentration	8.0 % by volume
product concentration	95.0 % by volume
reflux ratio	4.0
temperature of bottoms/live steam	107°C
temperature of feed	30°C
temperature of feed entering analyser	85°C
temperature of vapour leaving rectifier	78°C
temperature driving force in evaporator	3°C
Economic Variables :	
annual operational time	7200 h
life of equipment	10 yr
unit cost of electricity	1.25 Rs (kWh)$^{-1}$ (0.75 - 2.0)
unit cost of steam	250 Rs t^{-1} (150 - 300)
unit fixed capital cost of MVR system	0.15 m Rs (kl d^{-1})$^{-1}$ (0.075 - 0.2)

TABLE 1. PROCESS AND ECONOMIC VARIABLES

REFERENCES

Gundu Rao, G., (1964), "The overall steam requirement in an industrial alcohol distillery", Chem. Age of India, 16(1), pp 31-35.

Khattar, H.K., (Feb,1988), "Energy conservation with mechanical vapour recompression (MVR)", Alsthom International, India Liaison Office, New Delhi.

Manmohan and B.B. Paul, (1964), "Recovery of waste heat in power alcohol distillation plant", Chem. Age of India, 15(1), pp 12-16.

Panicker, P.K.N., (1964) "Vapor recompression system for steam economy in alcohol distillation plants", Chem. Age of India, 15(1), pp 36-41 .

Panicker, P.K.N., (August 1969), "Ethanol - Part IV, Distillation Technology", Chem. Processing & Engng, pp 27-32.

Paul, B.B. and Mitra, A.K., (1964), "Efficient use of waste heat in rectified spirit distillation plant", Chem. Age of India, 15(1), pp 9-11.

Venkiteswaran, S.L., (1964), "Operational efficiency in the alcohol industry in India - Part II", Chem. Age of India, 16(3), pp 230-235 .

Chapter 11

HIGH-TEMPERATURE PROCESS DRYING WITH HYBRID DESICCANT-HEAT PUMPS SYSTEMS

E Van den Bulck (Katholieke Universiteit Leuven, Belgium)

SUMMARY

This paper presents a hybrid drying system incorporating a rotary dehumidifier and a vapor compression heat pump. Whereas in conventional heat pump systems the drying temperature is the same as the condensor temperature of the heat pump, in the hybrid system, by operating the dehumidifier as a heat transformer, the drying temperature can be significantly higher than the condensor temperature. An energy performance index is defined as the ratio of the thermal energy use of a system equipped with a perfect recuperator to the primary energy use of the hybrid system. Results of a system simulation are presented showing this performance index as a function of drying and ambient conditions. The effect of several design parameters on the drying performance index is investigated and the optimal control of the system is discussed.

NOMENCLATURE

h_i	air enthalpy per unit mass of dry air at state (i)
N_{tu}	Number of transfer units for heat or mass transfer
P_{er}	Primary energy ratio or Performance factor defined by equation (4)
Q_{HT}	thermal energy use of air heater
Q_{DR}	reference energy use of ideal passive recuperative drying system
t_i	air temperature at state (i)
w_i	air absolute humidity at state (i)
W_{HP}	mechanical energy use of heat pump compressor
ϵ_{HX}	recovery heat exchanger effectiveness
λ	drying-chamber heat loss factor
ϕ_i	air relative humidity at state (i)

1 INTRODUCTION

Drying is an energy intensive process in which large amounts of sensible heat are required to vaporize the water condensed on or absorbed by the products in the drying chamber. Figure 1 illustrates an open drying system incorporating a heat exchanger to recuperate the sensible heat of the wet air stream exhausted from the drying chamber. The performance of this simple ventilation system can be increased significantly by installing a heat pump as shown in Figure 2. This heat pump reclaims the latent

Heat Pumps – Proceedings of the 4th International Conference, pp.127-134

heat of vaporization by condensing the water vapor contained in the exhaust of the drying chamber in the evaporator, and transferring this heat to the condensor where it is taken up by the ventilation air stream.

Open-cycle as well as closed-cycle heat pump systems for intermediate temperature drying applications have been widely accepted as cost effective (e.g., Oliver, 1982, Rao Pendyala et al., 1986). Dehumidifying heat pumps operating with R12 and R114 reach drying temperatures ranging from 25 to 75°C and 70 to 120°C , respectively, whereas heat pumps working with R22 are restricted to ± 50°C as maximum drying temperature (Krishna Kumar et al., 1984, Blaise, 1985). Primary energy savings up to 20% can be achieved depending upon the drying conditions. Large temperature lifts may require two-stage heat pump systems. Closed-cycle systems perform better than open-cycle systems, although these latter systems allow easier controls than the former systems. In view of the current CFC problem associated with R12 and R114, it is expected that these refrigerants will be gradually phased out in the near future.

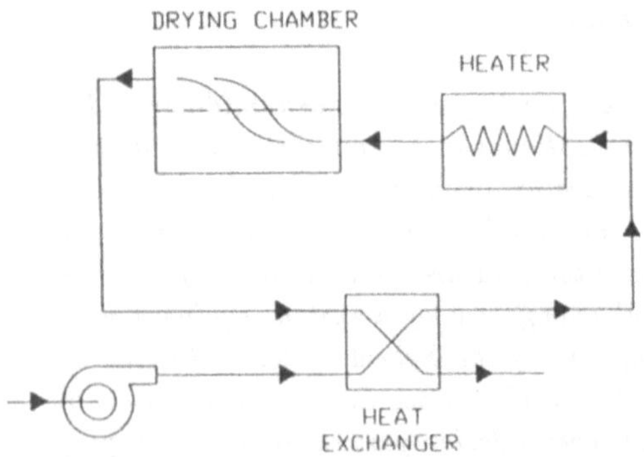

Figure 1: **Passive recovery ventilation drying system**

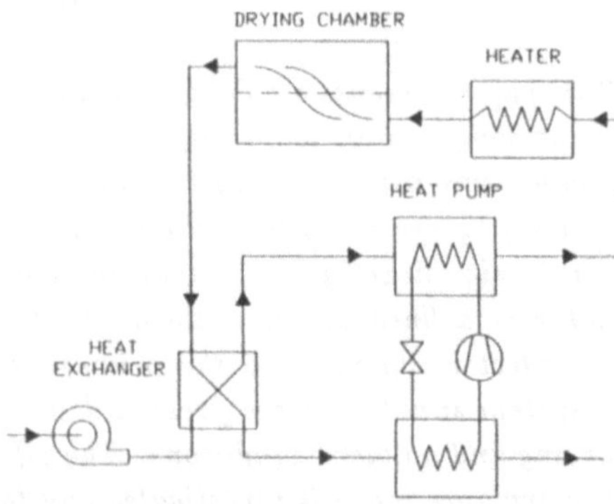

Figure 2: **Heat pump assisted ventilation drying system**

Rotary dehumidifiers are generally employed in air-conditioning applications where they are used for drying air streams at low temperature and are regenerated with hot air streams. These rotary dehumidifiers can also be operated as heat transformers when two air streams are used with equal temperatures and different humidities (Van den Bulck, 1987). In addition, these devices may be combined with heat pumps to form flexible hybrid air-conditioning systems (Burns et al., 1985). Open-cycle hybrid systems for drying applications can also be conceived and such a system is presented in this paper.

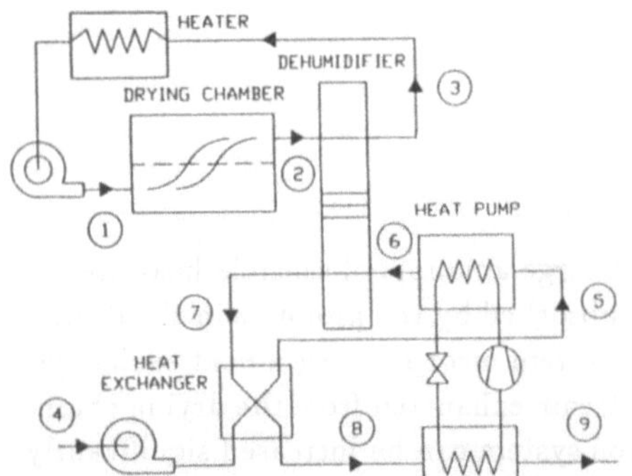

Figure 3: **Hybrid desiccant-heat pump drying system incorporating a rotary dehumidifier**

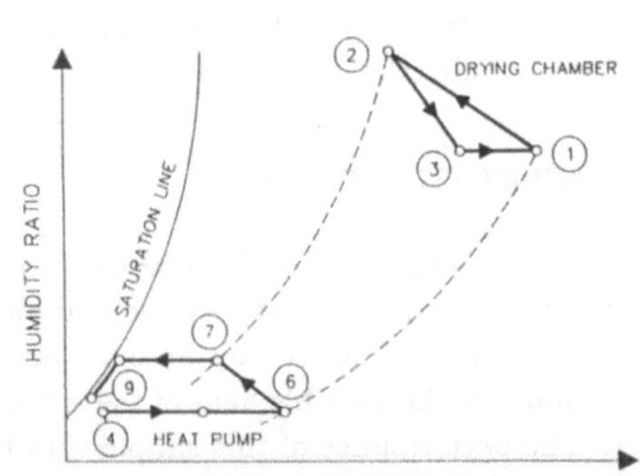

Figure 4: **Psychrometric processes of the hybrid drying system**

2 HYBRID DRYING SYSTEM

Figure 3 shows a schematic diagram of a hybrid drying installation with a rotary dehumidifier operating as a heat transformer. The psychrometric chart in Figure 4 illustrates the operating principle. Process air cycles through the drying chamber ($1 \rightarrow 2$), the rotary dehumidifier ($2 \rightarrow 3$), and the air heater ($3 \rightarrow 1$). Outside air is first passed through a recovery heat exchanger ($4 \rightarrow 5$) and then passed over the condensor of the heat pump ($5 \rightarrow 6$) where the temperature is increased to the required regeneration temperature. This air stream regenerates (dries) the desiccant wheel ($6 \rightarrow 7$), is passed through the recovery heat exchanger ($7 \rightarrow 8$) where most of the sensible heat is reclaimed and is then passed over the evaporator of the heat pump ($8 \rightarrow 9$) where the latent heat is reclaimed. The air stream may then be vented to the outside. The moisture removal rate and the overall performance of the system can be controlled by adjusting the dehumidifier wheel speed and the regeneration air flow rate.

2.1 Simulation

The system described in Figure 3 is simulated using a modular approach. The drying conditions of state (1) and the ambient conditions of state (4) are design conditions and are specified in each simulation. The air state properties at the other states, temperature t, absolute humidity w, enthalpy (per unit mass of dry air) h and relative humidity ϕ, are determined in a sequential order using separate unit models for each of the components. The air properties are correlated using standard expressions such that any pair of properties can be computed from the remaining pair (ASHRAE, 1989). The unit models can be described as follows.

Heat Exchanger. The performance of the heat exchanger is computed using an effectiveness approach with the heat exchanger effectiveness ϵ_{HX} as a parameter. Depending upon the inlet conditions, the humidity of air state 8 may be adjusted if t_8 falls below the dewpoint temperature of state 7. In this case condensation is already occurring in the recovery heat exchanger. Unless otherwise stated, a value of 0.9 is used for the heat exchanger effectiveness.

Dehumidifier. The outlet conditions of the rotary dehumdifier, states (3) and (7), are calculated from the inlet conditions (2) and (6), the wheel rotation speed and both air stream mass flow rates. Regular-density silica gel is used as the desiccant within the wheel. The model is based upon an ϵ-N_{tu} model for rotary heat and mass exchangers (Van den Bulck et al., 1985). Parameters are the overall N_{tu}'s for heat transfer and mass transfer with values used in the simulation of 6 and 3, respectively, and the thermal capacitance of the structural material used to construct the dehumidifier wheel. In the simulation, a value of 1500 J/°C /kg-dry-desiccant is used for this latter parameter unless otherwise specified.

Heat Pump. The heat pump is simulated using a thermodynamic model with 3°C subcooling at the outlet of the condensor, 2°C superheat at the inlet of the compressor and 70% isentropic efficiency of the compressor. Refrigerant 22 is used and it's properties are computed with the correlations developed by McMullan et al. (1985). The temperature differences between the air streams and the working fluid are assumed 4 °C and 8 °C in the condensor and evaporator, respectively. The temperature difference in the condensor is taken smaller because of the gliding air-stream temperature in that component. If state (9) is cooled below the dewpoint, the relative humidity is set equal to 1.0 and the air enthalpy and humidity are computed accordingly. The specific compressor work W_{HP} is calculated as

$$W_{HP} = (h_6 - h_5) - (h_8 - h_9) \quad (1)$$

Drying Chamber. The drying chamber is modeled as the adiabatic wetting of the air stream passing through, with a correction factor λ to account for the heat losses. The modeling equations are

$$t_2 = t_1 - \delta t$$
$$h_2 = (h_1 + \lambda h_{(t=t_2,w=w_1)})/(1+\lambda)$$

where δt is the temperature drop across the drying chamber. For this study, δt is taken as a design variable and set to 10°C . The simulation uses a value of 0.1 for λ.

Air Heater. If the temperature at the outlet of the dehumidifier t_3 is less than the specified drying temperature t_1, heat must be added to the ventilation air stream. The specific heat supply to the air heater, Q_{HT}, is computed from

$$Q_{HT} = h_1 - h_3. \quad (2)$$

2.2 Performance and Optimal Control

The performance of the system is described by the ratio of the primary energy use of an ideal recovery drying system as displayed in Figure 1 and the primary energy use of the hybrid drying system. The thermal energy use of the former system, Q_{DR}, is computed with the assumption that the recovery heat exchanger in Figure 1 has an effectiveness of 1:

$$Q_{DR} = h_1 - h_{(t=t_2,w=w_1)}. \quad (3)$$

The *Performance factor* or *Primary energy ratio* P_{er} is calculated by

$$P_{er} = \frac{Q_{DR}}{Q_{HT} + 3W_{HP}\dot{m}_4/\dot{m}_1}. \quad (4)$$

Defined in this way, the system is compared directly against an ideal passive recuperator system and the P_{er} is therefore less dependent upon the characteristics of the drying chamber. The numerical coefficient 3 in equation (4) reflects the relative higher price of electrical energy compared to thermal energy.

The system is controlled by adjusting the regeneration air flow rate $\dot{m}_4$ such that the outlet humidity of state (3) is equal to the specified humidity of the drying air at state (1):

$$w_3 = w_1.$$

In combination with a controlled heat supply to the air heater, the air stream cycling through the drying chamber and dehumifier wheel is then in steady state. The control of the power supply to the heat pump is simulated by varying the outlet temperature of the heat pump t_6. This temperature and also the dehumidifier wheel speed are adjusted such that the performance index P_{er} is maximized.

3 PERFORMANCE OF HYBRID DRYING SYSTEM

Modular simulation of the units, controls and optimization procedures as explained in the previous paragraph leads to the following general results.

- The optimal installed compressor capacity relative to the reference energy use of the drying system, W_{HP}/Q_{DR}, is small and ranges from about 8 to 15%, depending upon the operating conditions.
- The ratio of the optimal regeneration flow rate to the ventilation flow rate, $\dot{m}_4/\dot{m}_1$ varies widely as the operating conditions vary. Optimal ratios of 0.25 to 4 may result.

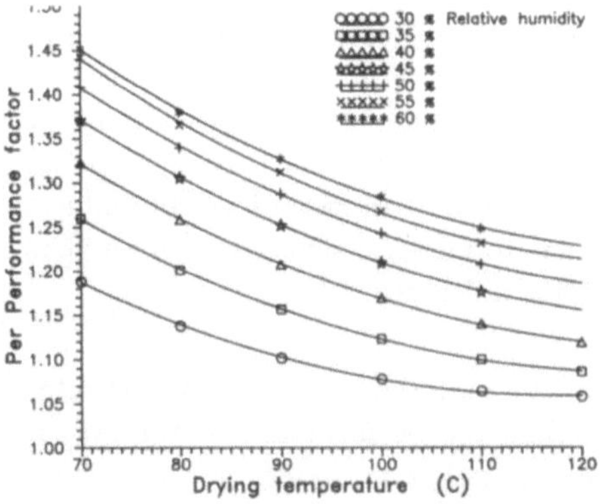

Figure 5: P_{er} **of hybrid system as a function of drying temperature** t_1 **and relative humidity** ϕ_1**. Ambient conditions:** $t_4 = 12°C$ **and** $\phi_4 = 0.8$**.**

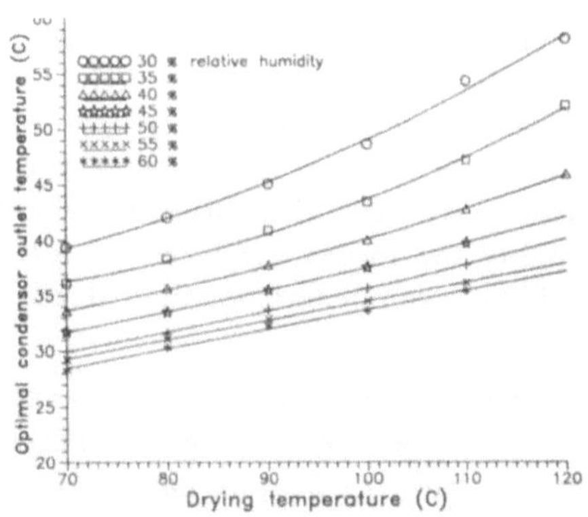

Figure 6: **Optimal condensor outlet temperature in the hybrid system as a function of drying conditions. Ambient conditions as specified on the left.**

- The optimal wheel speed varies in a range from 1 to 6 depending upon the operating conditions. In drying applications the dehumidifier wheel spins about 5 to 10 times faster than in air-conditioning applications.

Actual hybrid systems should include adaptive controls for the air heater, the wheel rotation speed and both air blowers.

The performance of the hybrid drying system for standard ambient conditions is shown in Figures 5 and 6. Figure 5 shows the P_{er} of the drying system as a function of the drying temperature t_1 with the relative humidity at the entrance of the drying chamber ϕ_1 as a parameter. Similar to conventional heat pump systems, the performance of the hybrid system drops with increasing drying temperature and decreasing drying hymidity. However, for products which require drying temperatures above 100°C , steam-air mixtures may be used with increased humidity (± 0.4 kg-water/kg-dry-air) which leads to a reduction in drying time (Chin Chu et al., 1959). In the high temperature range, relative humidities of 50 to 60% are common practice and the performance of the hybrid drying system is then accordingly high.

The optimal condensor outlet temperature t_6 is shown in Figure 6 as a function of drying temperature and humidity. This temperature increases with increasing drying temperature and decreasing drying humidity. As the drying state (1) moves downwards and to the right in the psychrometric diagram of Figure 4, the regeneration state (6) has to shift correspondingly towards higher temperatures in order for the dehumidifier wheel to perform as a heat transformer. Figure 6 indicates that condensor temperatures well below 55°C suffice to yield drying temperatures in the range of 70 to 120°C , and that R22 heat pumps can thus be used in this temperature range.

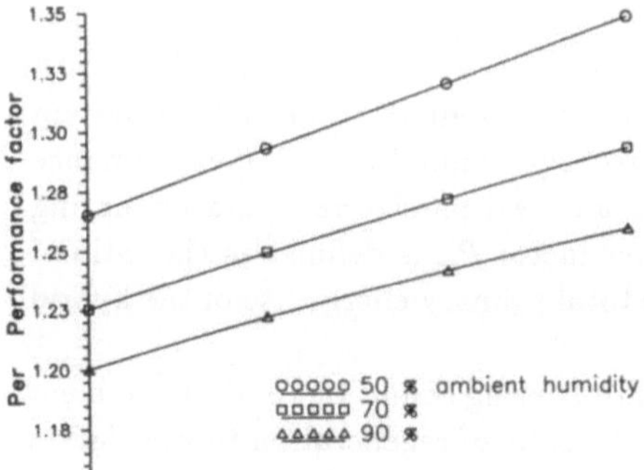

Figure 7: P_{er} **of hybrid system as a function of ambient temperature** t_4 **and relative humidity** ϕ_4**. Drying conditions:** $t_1 = 100°C$ **and** $\phi_1 = 0.5$**.**

Figure 7 shows the change of the P_{er} performance factor with ambient conditions. The performance of the hybrid system increases with ambient temperature and decreasing humidity. This latter effect can be explained by the fact that the temperature lift of the rotary dehumidifier $t_3 - t_2$ increases with the driving humidity difference $w_2 - w_6$, where $w_6 = w_4$ (Van den Bulck, 1987). In case of a high ambient humidity, the optimal condensor temperature has to increase strongly with ambient humidity to compensate for the reduced humidity difference.

The effect of some system parameters on the performance factor is illustrated in Figures 8 and 9. Figure 8 demonstrates that the performance of the system increases with the heat exchanger effectiveness and that the effect is pronounced. High performance recovery heat exchangers are required in order for the hybrid drying system to perform well. The optimal condensor temperature is almost independent of the effectiveness of the heat exchanger, as is the optimal installed compressor capacity and the optimal wheel speed. However, the optimal regeneration flow rate increases strongly with ϵ_{HX}.

Figure 9 demonstrates that the performance of the hybrid system decreases with increasing dehumidifier-matrix thermal capacitance. The optimal condensor temperature increases with this parameter, while other variables such as wheel speed may also change. This effect is explained by the fact that the rotary dehumidifier '*loses*' more thermal energy from the high-temperature ventilation air to the intermediate temperature regeneration air as the thermal capacity increases. Low capacitance, silica gel dehumidifier wheels are therefore needed for these applications.

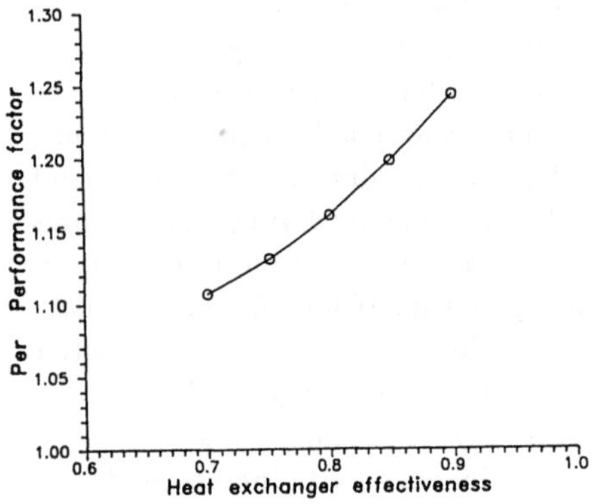

Figure 8: P_{er} **of hybrid drying system as a function of recovery heat exchanger effectiveness. Drying and ambient conditions as previously specified.**

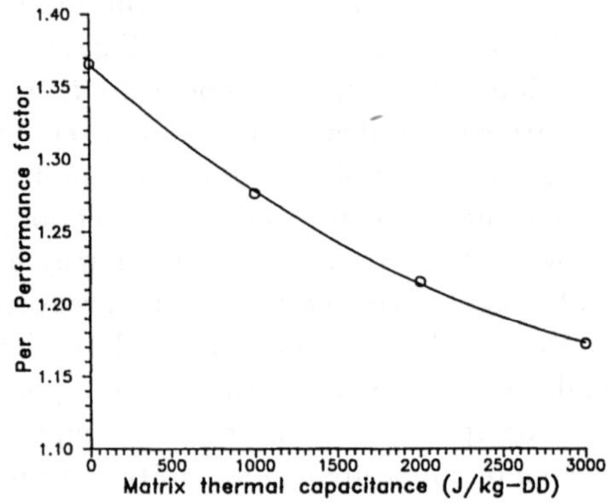

Figure 9: P_{er} **of hybrid drying system as a function of matrix thermal capacitance (per unit dry desiccant).**

4 CONCLUSION

A theoretical study of a hybrid drying system is presented. The hybrid system incorporates a R22 heat pump and a silica gel rotary dehumidifier. The performance of the hybrid system is studied in reference to an ideal passive recuperative drying system. A Primary energy ratio or Performance factor P_{er} is defined as the ratio of the energy use of this reference system and the total primary energy use of the hybrid system.

This P_{er} has been computed as a function of the drying conditions and the ambient conditions. The installed heat pump capacity, the ratio of regeneration to ventilation flow rate and the dehumidifier wheel speed are optimized to yield maximal performance. The P_{er} ranges from 1.10 to 1.45 for well-operated systems and for drying temperatures ranging from 70 to 120°C . Depending upon the ambient and drying conditions, primary energy savings with the hybrid system compared to conventional fuel-fired systems equipped with a heat recovery system can be as high as 40%.

The hybrid system extends the operating range of R22 heat pumps. The analysis shows that drying temperatures well above 100°C are possible for condensor temperatures ranging from 30 to 50°C . In view of the current CFC problem associated with R114, R22 units can thus be used in the temperature ranges previously accessible only by R114 devices.

REFERENCES

ASHRAE, 1989, *Pyschrometrics*, ASHRAE Handbook of Fundamentals, Ch. 6, ASHRAE Inc., Atlanta, U.S.A.

Blaise J.C., 1985, "Pompe a Chaleur Haute Temperature et Applications Industrielles", *Revue Generale du Froid*, Vol. 75, No. 11, pp. 643-648.

Burns, P.R., J.W. Mitchell and W.A. Beckman, 1985, "Hybrid Desiccant Cooling Systems in Supermarket Applications," *ASHRAE Trans.*, Vol. 91, Pt 4., pp. 457-468.

Chin Chu, J., S. Finelt, W. Hoerrner and Min-Shuey Lin, 1959, "Drying with Superheated Steam-Air Mixtures", *Industrial and Engineering Chemistry*, Vol. 51, No. 3, pp. 275-281.

Krishna Kumar D., S. Srinivasa Murthy and M.V. Krishna Murthy, 1984, "Studies on Air-to-Air Heat Pump for Drying Applications", *Heat Recovery Systems*, Vol. 4, No. 4., pp. 209-217.

McMullan, J.T., D.W. Hughes and R. Morgan, 1985, "A Suite of Computer Programs for Calculating Refrigerant Properties", *Heat Recovery Systems*, Vol. 5, No. 2, pp. 143-180.

Oliver T.N., 1982, "Process Drying with a Dehumidifying Heat Pump", *Proc. Int. Symp. Industrial Applications of Heat Pumps*, BHRA Fluid Engineering, Cranfield, U.K., pp. 73-88.

Rao Pendyala V., S. Devotta and V.S. Patwardhan, 1986, "The Economics of Heat Pump Assisted Drying Systems" *Heat Recovery Systems*, Vol. 6, No. 6, pp. 432-442.

Van den Bulck, E., 1988, "Performance Characteristics of Open-Cycle Solid Desiccant Heat Transformers", *Proc. 2nd Int. Workshop on Research Activities on Advanced Heat Pumps*, pp. 69-77, Graz, Austria.

Van den Bulck E., J.W. Mitchell and S.A. Klein, 1985, "Design Theory for Rotary Heat and Mass Exchangers", *Int. J. Heat Mass Transfer*, Vol. 28, No. 8, pp. 1575-1595.

Chapter 12

HEAT PUMPS FOR HEATING AND COOLING OF HOTELS

G Eggen (SINTEF Refrigeration Engineering, Norway)

Summary

Hotels usually have cooling demands which require the installation of a refrigeration plant. At the same time, heating is required. For this reason, heat pumps for combined heating and cooling purposes in hotels can be very efficient and economical.

This paper describes heat pump plants for heating and cooling of three different hotels in Norway. The heat pumps were mainly installed to cover cooling demands in summer. However, they were also able to cover most of the hotels' heat demands. Throughout the year, there is a constant demand for hot tap water which can also mean long operation times and favourable conditions for heat pumps. In each of the heat pump systems described, the payback time was less than two years.

HEAT PUMPS FOR HEATING AND COOLING OF HOTELS

1. INTRODUCTION

During recent years, the insulation of buildings has been improved, while at the same time the amount of heat coming from the electric equipment (lighting, computers etc.) has increased. This means that the heating demand has decreased considerably. On the other hand this has created a cooling demand in buildings.

The cooling demands of hotels are such that they usually require the installation of a refrigeration plant. At the same time, heating is also required, especially for service hot water. If a refrigeration plant for cooling purposes is installed, it is possible to use it as a heat pump during the heating season. Therefore, heat pumps for combined heating and cooling purposes can be very profitable in hotels.

In this paper, heat pump plants for heating and cooling of three different hotels in Norway are presented.

Heat Pumps – Proceedings of the 4th International Conference, pp.135-142

2. DESCRIPTION OF THREE DIFFERENT HOTELS WITH HEAT PUMPS

Table 1 gives some characteristic data from the three hotels with heat pump systems.

Plant	Hotel Maritim Haugesund	Alexandra Hotel Loen	Royal Garden Trondheim
Building			
Total floor area (m^2)	13.500	16.000	23.000
Conference hall area (m^2)	1.200	1.200	1.400
Number of rooms (-)	160	200	300
Number of restaurants(-)	2	4	6
Swimming pool	No	Yes	Yes
Heat pump			
Heating capacity (kW)	300	400	900
Refrigerant	R12	R500	R22
Heat source/heat sink	sea water	exhaust air	river water

Table 1. Three hotels with heat pumps for heating and cooling

3. HOTEL MARITIM, HAUGESUND

In 1985 the old hotel building of Hotel Maritim in Haugesund, Norway, was reconstructed. A new congress hall was built in the annexe. In connection with this expansion a heat pump for combined heating- and cooling purposes was installed (1).

3.1 Heat Pump Capacity

The heat pump is sized to cover the cooling demand in the new hotel building, which is 225 kW at 6 ° C cooling water temperature out of the evaporator. With 50°C forward water flow temperature out of the condenser, the condenser heat production is 310 kW. This covers most of the heat demand in both the new and existing buildings of the hotel.

3.2 Plant Design

A principal piping diagram of the heat pump for heating and cooling is shown in Fig. 1. The heat pump is operating with R12 as working fluid. On the low temperature side, the heat pump extracts heat from a water/glycol system, which is connected to the air cooling heat exchangers in the hotel and to a titanium sea water plate heat exchanger.

The heat pump unit is located on the quay just outside the hotel. The sea water system is quite simple, consisting of a 4 m pipeline with a submersible pump in a basin underneath the heat pump. The pump basin is directly connected to the sea through a perforated plate acting as strainer. After passing through the plate heat exchanger, the sea water is drained back to the sea.

The heat pump is connected to the heating system in the old building by means of insulated pipes. The return water is preheated by the heat pump and at low ambient temperature, it will be further heated by either an electric or an oil fired boiler. The heat distribution system consists of a radiator system and heat exchangers in the ventilation system. Service hot water is heated by the heat pump in separate hot water tanks (4000 liters). To produce high temperature hot water for kitchen purposes, some of the preheated water flows to an electric heated water tank, where the water is heated up to the desired temperature.

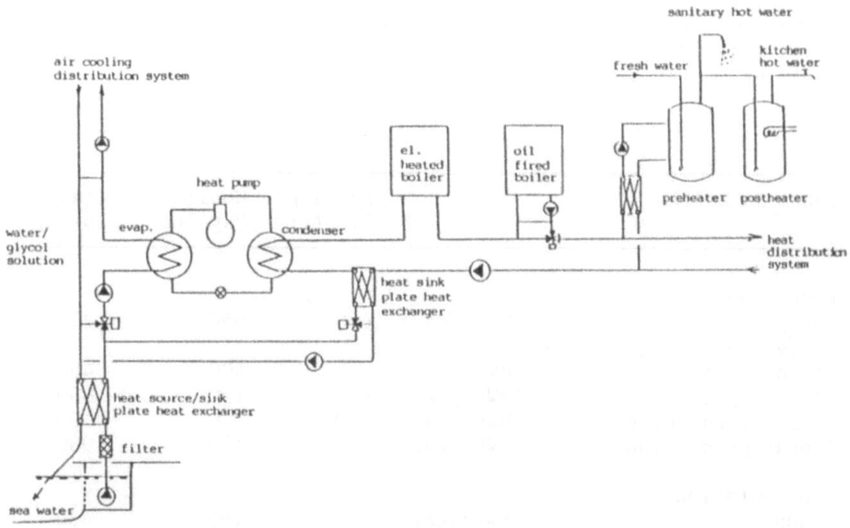

Fig. 1. Principal piping diagram for the heat pump at Hotel Maritim

At the design outside temperature of -10 ° C, the water flow temperature for heat distribution to the old part of the hotel needs to be 70°C. The required water temperature decreases with increasing ambient temperatures. However, the flow temperature has a lower limit of 50 ° C, in order to heat the service water to at least 45°C. Typically, in 1989 the service hot water consumption was 7.100m^3. The average heat demand for preheating this throughout the year was 40 kW.

The heat pump normally transfers heat from the air cooling heat exchangers to the heat distribution system. When the heat requirement dominates, the heat from the air cooling system is not sufficient and the heat pump will use the sea water as an additional heat source.

In summertime, when the cooling requirements are dominant, the sea water will then be used as heat sink for the air cooling plant. The surplus heat is transferred from the heat distribution system to the sea water via the two heat exchangers shown in Fig. 1. The operation of the heat pump for both heating and cooling is handled by an automatic control system.

3.3 Operating Experience

The heat pump has been in operation for more than four years. During the first year, there were some problems with the automatic control equipment, which resulted in compressor damage. The compressor was replaced in 1986 and during 1987 and 1988, the availability was close to 100%. However, the control problems were not fully solved until 1989, after a second compressor breakdown.

Normally, cooling requirements dominate from April to September. The rest of the year, the plant operates primarily as a heat pump. Fig. 2 shows measurements of weekly mean values of heat production from August 1987 to August 1988. During the summertime (April to September) the weekly mean cooling load was approximately 65% of the measured condenser heat production, with most of this heat being dumped into the sea.

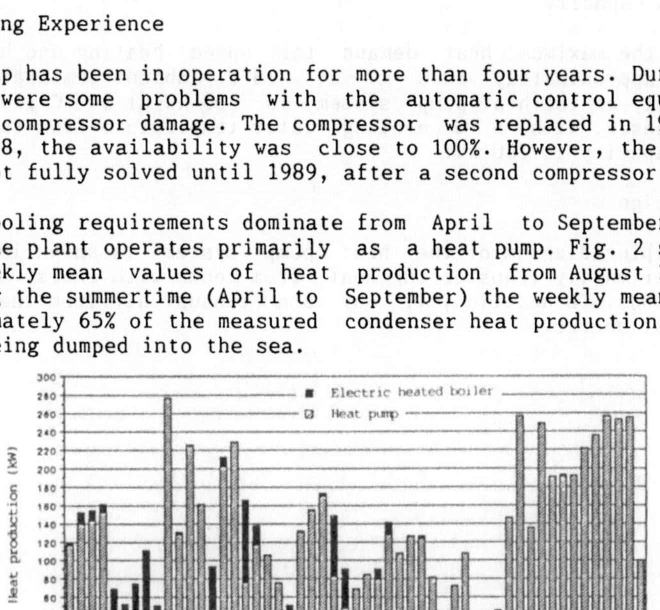

Fig. 2. Weekly mean values of heat production from heat pump and electric heated boiler from August 1987 to August 1988

Table 3 gives some results from energy measurements during one year of operation. Peak heating from the electric heated boiler could have been reduced considerably by better control equipment.

Operation		Heating Sep-Apr	Cooling May-Aug	Total Sep-Sep
Heat production				
Heat pump	(MWh/year)	650	550	1.200
Heat rejected during cooling operation (estimated)	(MWh/year)		400	400
Useful heat prod	(MWh/year)	650	150	800
Electric heated boiler	(MWh/year)	100	0	100
Total heat production	(MWh/year)	750	150	900
Energy consumtption				
Heat pump	(MWh/year)	235	195	430
Electric heated boiler	(MWh/year)	100	0	100
Total energy consumption	(MWh/year)	335	195	530
Energy savings	(MWh/year)	415	150	565

Table 2. Energy measurements from September 1987 to September 1988

The additional expence of converting the refrigeration plant for heat pump operation, was approximately NOK 300.000,- (1985). The payback period for this extra investment is less than two years compared with electric heating (electricity price 0.4 NOK/kWh), and less than four years compared with an oil fired boiler (oil price 2.0 NOK/liters).

4. ALEXANDRA HOTEL, LOEN

This heat pump system was installed late 1987, and it has been in operation since January 1988.

4.1 Heat Pump Capacity

In this hotel the maximum heat demand for space heating and hot tap water production is approximately 1000 kW at -13°C ambient temperature (2). The heating capacity of the heat pump system is 400 kW at 54 ° C flow temperature from the condenser. With 6 ° C cooling water temperature from the evaporator, the cooling capacity is 280 kW.

4.2 Plant Design

A principal piping diagram of the heat pump system is shown in Fig. 3. The heat pump is primarily transferring heat from rooms with excess heat to rooms with heat demands. In addition, it may use exhaust air from the hotel as an additional heat source.

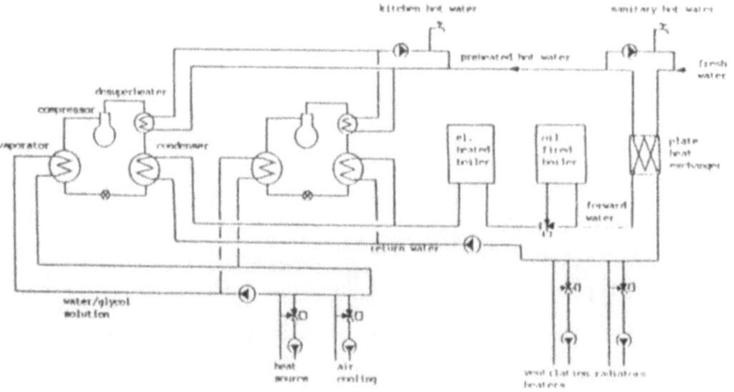

Fig. 3. Principal piping diagram for the heat pump at Alexandra Hotel

The heat pump system consists of two heat pumps which are working in parallell on both the low temperature side and the high temperature side. The heat pumps extract heat from a water/glycol circuit which is connected to air coolers in the restaurants, the laundry and the swimming pool and to heat exchangers in the exhaust air channels.

The heat pumps are connected to the heat distribution system to preheat the return water before it is heated by the electric heated and oil fired boilers. The heat distribution system consists of radiators in different rooms and heating coils in the ventilation systems.

The capacity is controlled by a thermostat in the flow water to keep this temperature at a constant level (54 °C). Further heating is added by the electric heated boiler when the flow water temperature decreases below 52°C.

Hot tap water is preheated to approximately 50°C by heat exchange with the flow water from the heating system in two plate heat exchangers. Service hot water at 50 °C is distributed to the hotel rooms, while hot water for kitchen purposes is further heated to 70 - 80°C by cooling the superheated refrigerant gas from the compressors.

The hot tap water system has no storage (hot water tanks) except the volume of the piping system for hot water circulation. The plate heat exchangers are therefore dimensioned to cover the maximum heat demand for service hot water at 50°C. In 1989, the total hot water consumption at Hotel Alexandra was approximately 9.500 m^3 and the average heat demand for hot water production throughout the year was 70 kW. The accumulation curve for hot water during a typical day is shown in Fig. 4.

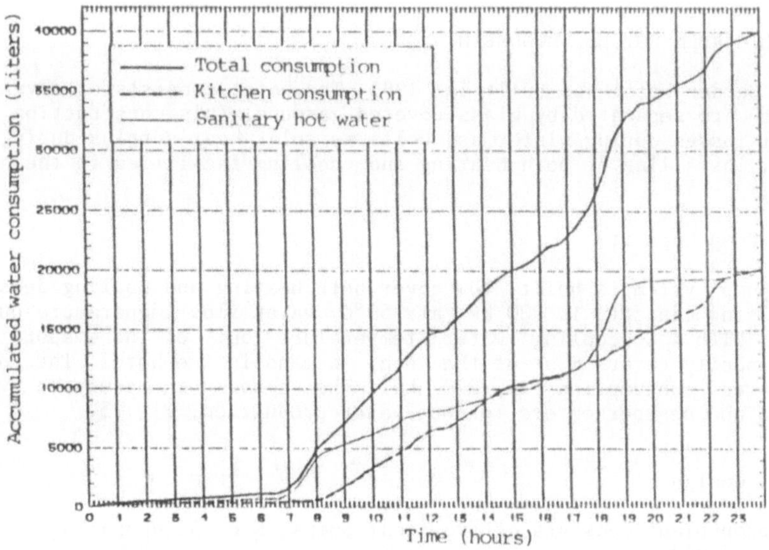

Fig. 4. Hot water accumulation curves from September 29., 1989.

4.3 Operating Experience

The heat pumps were started up in January 1988, and have been in operation for more than two years without any problems with the exception of one compressor replacement due to manufacturing fault (January 1988). Energy measurements from 1989 shows that their average heat production has been approximately 50%, of maximum capacity. Measurements from 1987 and 1988 show very clearly how they reduced the energy consumption for the hotel, Table 3.

Year		1987	1988
Heat production			
Heat pump (estimated)	(MWh/year)	0	2000
Electric heated boiler	(MWh/year)	230	50
Oil fired boiler (η=0.7)	(MWh/year)	1960	400
Total heat production	(MWh/year)	2190	2450
Energy consumption			
Heat pump	(MWh/year)	0	695
Electric heated boiler	(MWh/year)	230	50
Sum electricity - heating	(MWh/year)	230	745
Oil fired boiler	(MWh/year)	2800	580
Sum energy - heating	(MWh/year)	3030	1325
Other electricity	(MWh/year)	3070	2930
Total energy - hotel	(MWh/year)	6100	4255

Table 3. Energy measurements from two years, with and without heat pump at Alexandra Hotel

The total investment for the heat pump installation was NOK 1.400.000,- (January 1988). The heat pump installation was an alternative to using a dehumidifier for the swimming pool, an air cooler for the dining hall and sound insulating an old chiller on the roof. These alternative investments would have cost NOK 600.000,-, therefore the expense for the heat pump installation was NOK 800.000,-. However in 1988 the energy costs at Alexandra Hotel were reduced by NOK 500.000,- compared to 1987. Here the payback period for the heat pump system was less than two years.

5. ROYAL GARDEN HOTEL, TRONDHEIM

The Royal Garden Hotel was built in 1983. The hotel consists of several buildings which are separated by glass covered gardens. This construction involves great heat losses during winter as well as solar heat surplus during summer. Therefore, installing of both heating and cooling facilities in the hotel was necessary.

5.1 Heat Pump Capacity

The heat pump system is built to cover both heating and cooling demands. The design heating capacity is 900 kW at 50°C water flow temperature out of the condencer, with 4°C cooling water temperature out of the evaporator. This heating capacity covers most of the heat demand in the hotel. The compressor design power consumption is 220 kW. The system is equipped with both subcoolers and desuperheaters for hot water production, Fig. 5.

5.2 Plant design

The heat pump plant consists of two heat pumps, Fig. 5, operating with R22 as the working fluid.

The hotel is located on the riverside at the mouth of the river Nidelven and brackish water from the bottom of the river is used as an additional heat source.

The heat pumps extract heat from a water/glycol solution, circulating in a pipeline which is connected to both air coolers and cold storage chambers in the hotel as well as to a river water plate heat exchanger. The brackish water mainly consists of sea water, which is streaming from the Trondheim fjord into the bottom of the river mouth, as fresh water is flowing out at the top of the river. This brackish water temperature is rarely more than 10°C in the summer, Fig. 6, and since the air coolers in the hotel are sized for water temperatures of 12 ° C, the cooling demand is usually covered by this plate heat exchanger. In fact, the heat sink plate heat exchanger which should be used to dump excessive heat from the heat pumps, has only operated a few hours every year.

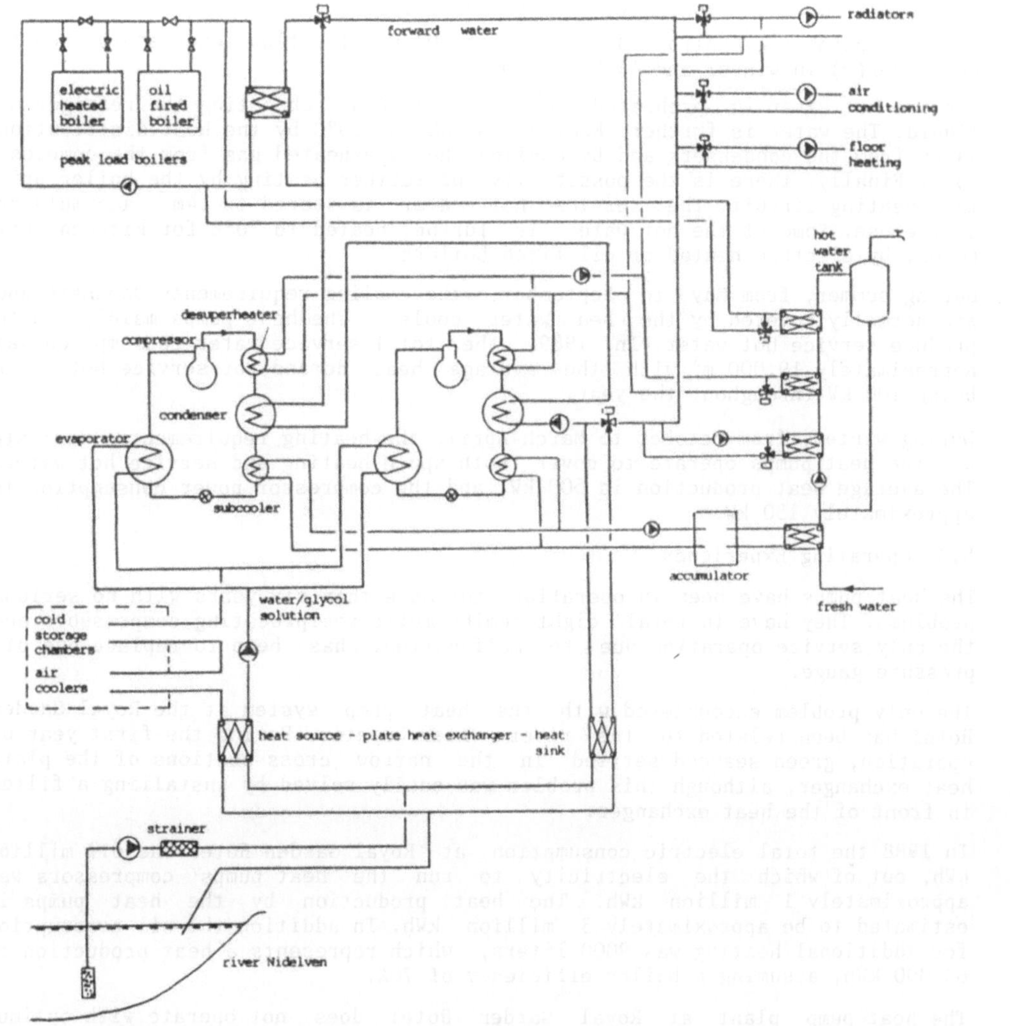

Fig. 5. Principal piping diagram for the heat pump plant at Royal Garden Hotel

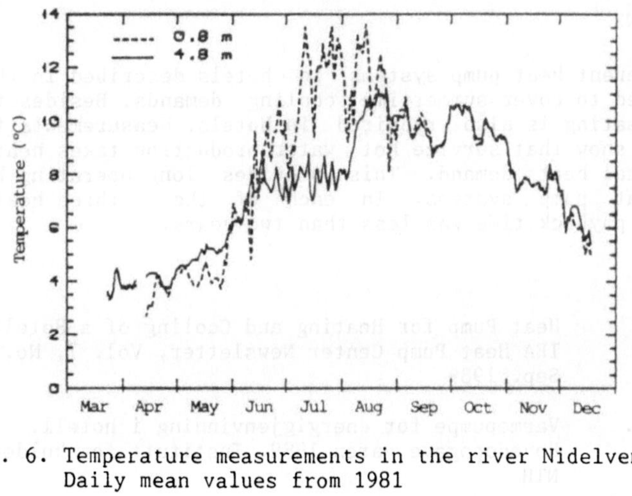

Fig. 6. Temperature measurements in the river Nidelven
Daily mean values from 1981

On the high temperature side, the condensers of the two heat pumps are connected in series in the return water line upstream of the electric heated and oil fired boilers. The heat distribution system consists of radiators, a floor heating system and heaters in the ventilation system.

The heat pumps are controlled to produce 50 ° C water flow temperature from the condenser(s) in winter and 46°C in summer.

Service hot water is preheated to 25 - 30°C by subcooling the refrigerant liquid. The water is further heated to 45 - 50°C by the heat distribution water from the condensers and by cooling the superheated gas from the compressors. Finally, there is the possibility of further heating by the boiler primary heating circuit. The service hot water is stored in $14 m^3$ accumulators before use. Some of the hot water is further heated to 70°C for kitchen purposes, by electric heated or oil fired boilers.

During summer, from May to September, the cooling requirements dominate and are normally covered by the sea water cooler. The heat pumps mainly run to produce service hot water. In 1989, the total service water consumption was approximately 19.000 m^3 with the average heat demand for service hot water being 100 kW throughout the year.

During winter, from October to March-April, the heating requirements dominate and the heat pumps operate to cover both space heating and service hot water. The average heat production is 500 kW, and the compressor power consumption is approximately 150 kW.

5.3 Operating Experiencs

The heat pumps have been in operation for more than six years with no serious problems. They have in total eight semihermetic reciprocating compressors and the only service operation due to malfunction, has been to replace one oil pressure gauge.

The only problem encountered with the heat pump system at the Royal Garden Hotel has been related to the river water system. During the first year of operation, green seaweed settled in the narrow cross-sections of the plate heat exchanger, although this problem was easily solved by installing a filter in front of the heat exchanger.

In 1988 the total electric consumption at Royal Garden Hotel was 6.7 million kWh, out of which the electricity to run the heat pumps' compressors was approximately 1 million kWh. The heat production by the heat pumps is estimated to be approximately 3 million kWh. In addition the oil consumption for additional heating was 9000 liters, which represents a heat production of 63.000 kWh, assuming a boiler efficiency of 70%.

The heat pump plant at Royal Garden Hotel does not operate with optimum efficiency on the hot water production system. The coefficient of performance may be improved by direct cooling of refrigerant liquid using fresh water instead of indirect cooling using a water circuit with accumulator (20 m^3) as shown in Fig. 5. In spite of this, the heat pumps have proved to be very successful. In 1983 the investment for the two heat pumps was NOK 765.000,-. The payback period was less than two years.

6. CONCLUSIONS

The three different heat pump systems in hotels described in this paper were mainly installed to cover summertime cooling demands. Besides the demand for air cooling, heating is also required in hotels. Measurements from the three example hotels show that service hot water production takes nearly fifty percent of the total heat demand. This provides long operating hours and good economy for heat pump systems. In each of the three heat pump systems described, the payback time was less than two years.

REFERENCES

1. Eggen, G. Heat Pump for Heating and Cooling of a Hotel. IEA Heat Pump Center Newsletter, Vol. 7, No. 3, Sep. 1989

2. Jahren, E. Varmepumpe for energigjenvinning i hotell. Hovedoppgave våren 1989. Institutt for kuldeteknikk, NTH

Chapter 13

THERMODYNAMIC GESTION AND THERMAL COMFORT IN A SOLAR TV COMPLEX BUILDING IN BRUSSELS: FIRST RESULTS

A Gillet (Atelier 'D' SC Wavre, Belgium)

SUMMARY

Based on previous experience (GILLET 1985-1987), a monovalent heat pump , refrigeration system has been designed and set up to fullfill the heating and cooling needs of a big TV complex building now completed in Brussels.
The originality of the system lies in independant heat sinks and sources linked through a rain water cistern, namely
- Water pipe net buried under the building
- Water pipe net buried in the ground around the building
- Static refrigerant heat exchangers either submerged in the cistern or hanging from the ceiling of the studios for retrieval of waste heat emited by the lighting.
- Dynamic refrigerant heat exchangers working either on the cistern water or the ventilation air.
A huge roof integrated solar collector is part of the system, linked with the air ducts for preheating of air and with the cistern for extra heat collection.
The system is controlled and monitored by an automtic recording programmable device.
The results recorded during the first winter month of operation are presented. Thermal comfort requirements are met with a reasonable electricity consumption.

I INTRODUCTION

The building itself (fig 1) is caracterized by an uptodate architecture integrating the more recent experience in bioclimatics :

- buried construction for a large proportion
- high form coefficient : 6,44 m3/m2 of envelope
- thermal insulation particularly good : 0,36 W/m2.k.
- Square base pyramidal roof the diagonal of which is oriented south-north.

Two faces, south east and south west, are treated as bivalent solar collectors. The air gap between single glass cover and one third translucent back insulation is used for preheating of ventilation air. The structure itself is hollow and finned to ensure the heating of water fed through it on a "drain down" basis.
Two faces, north-east and north-west present a gap, between metallic roofing and insulation, through which air can be sucked, during summer nights, for free cooling.
- controled mechanical ventilation limited to hygienic needs (0,7 vol/h) with heat pipes exchangers to retrieve heat from exhaust air.

The thermodynamic system itself includes two separate units (fig 2)

Heat Pumps – Proceedings of the 4th International Conference, pp.143-148

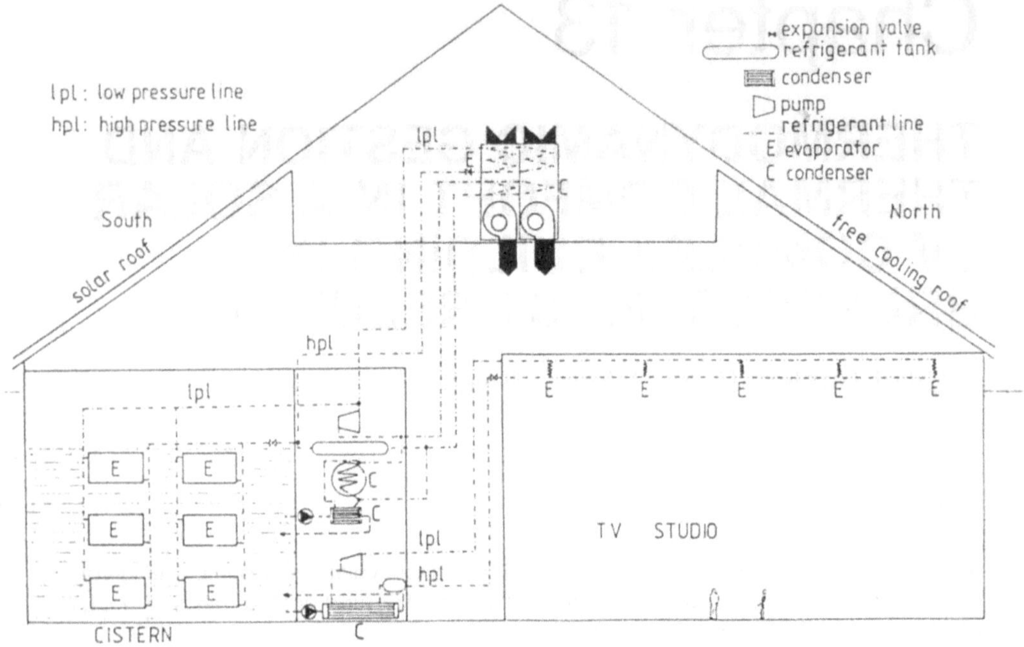

Fig 1 : General schematic layout of building and appliances

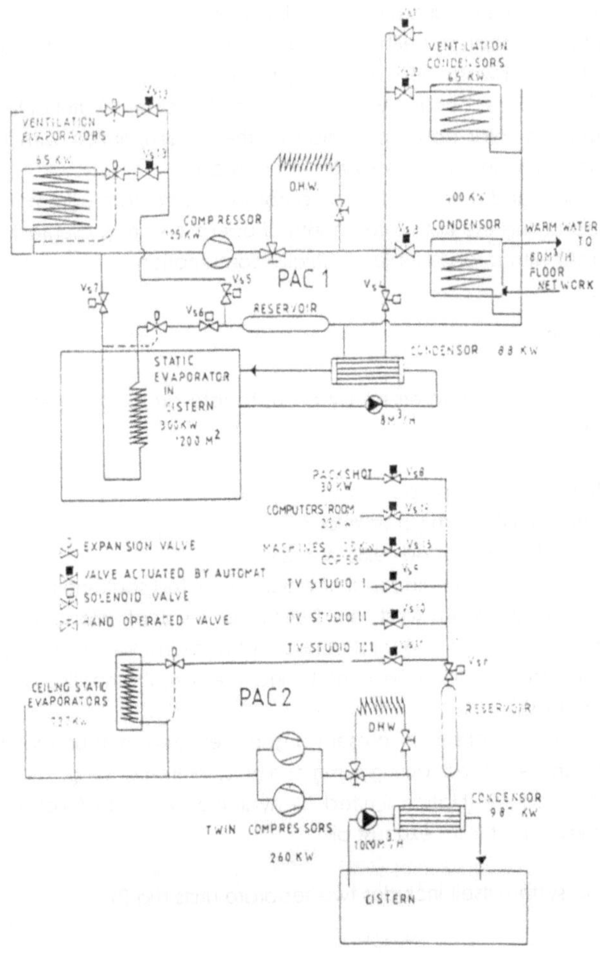

Fig 2 : Schema of refrigerant networks

PAC 1

aimed chiefly at calorific production, it heats loop water circulated in the concrete floors of the building and heat or cools ventilation air. It takes low temperature energy from a 150 m3 rain water cistern through static evaporators and rejects it through a dynamic condensor.

PAC 2

aimed only at refrigeration and retrieval of the internal heat emited by the lighting of studios and by appliances.It is equiped with static evaporators hanging from the ceiling of the rooms to be controlled and with a dynamic condensor rejecting heat in the cistern water.

Compensation and storage of heat by the cistern water is completed by piping nets through which water can be circulated as needed and which are laid :

- under the building itself
- in the earth around the building, moved during excavation
- in the earth 1,5 m deep under the parking lots.

Domestic hot water is prepared by exchange with refrigerant in derivation circuits between compressors and condensors (desuperheating).

II CONTROL METHODOLOGY

A control and data acquisition system is provided aimed at insuring thermal comfort according to the bioclimatic character of the premises with as little as possible energy consumption.

The headlines of the control program are given below

Taking advantage of the heavy character of the concrete floors, heating is chiefly provided during off-peak hours.

To avoid overheating by direct sun rays, set temperature are lower during day time than during night and the night operation of PAC 1 is limited in time according to outside temperature.

Heat loading of concrete floors is thus inversely proportionnal to anticipated heat needs of the next day.

- Refreshment of premisses and retrieval of lighting heat is controlled by underceiling temperatures and /or relative humidity at predetermined levels.
- Cistern water temperature fluctuations are limited between 4 and 20°c, by connecting successively buried piping nets and pumping water through them either to store excess heat in soil or to retrieve heat stored as needed.

- Ventilation is carried according to temperature reached at the head of solar collectors.

by internal recycling eventually, with thermodynamic heating, or

through the air gaps of the collectors or

directly by exterior air to avoid overheating. In the last circumstance cistern water is pumped through the solar collector structure for heat retrieval

- Ventilation is stopped during off office hours, except during hot summer when free cooling of the building fabrics is operated.

III RESULTS

The owner has moved in the premises in the beginning of september 1989, even before heat pumps have been electrically connected.

Operation of PAC 1 began in mid october without any regulation on an on-off basis. Anyhow the bioclimatic character of the building has provided a relative thermal comfort to the inoccupants during the setting up of controls and the test period which extended itself until the end of november.

Recording printer, ready by mid november, has registered until now the mean parameters of the building and the appliances. Data recorded are summarized in diagrams n°1 and 2 appended. They need the following comments.

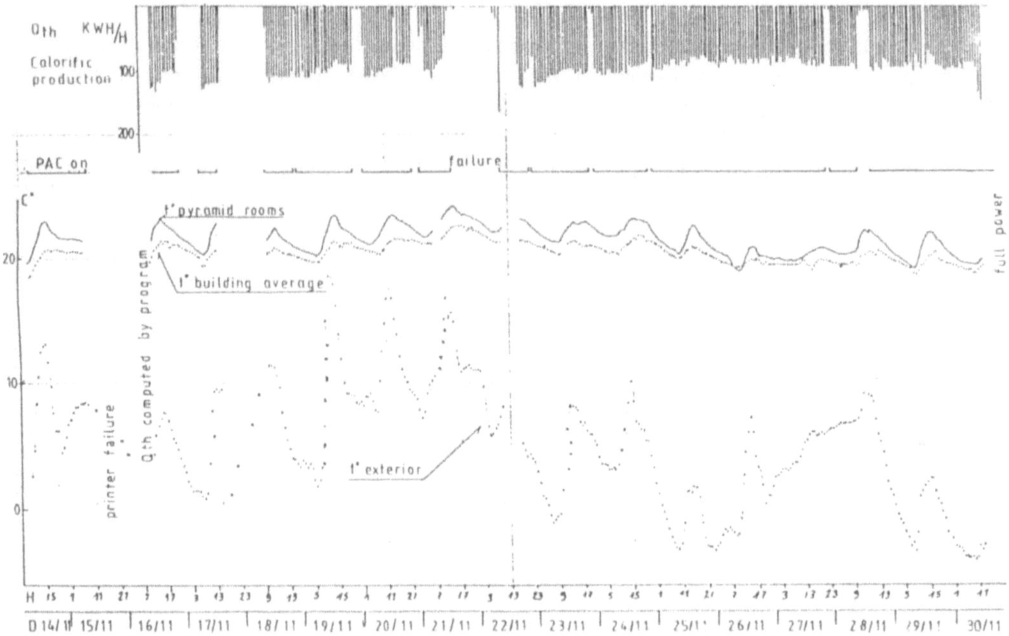

Diagram n° 1 : Average hourly temperatures and calorific production during tests out setting and period (NB PAC 1 locked at 25 % power)

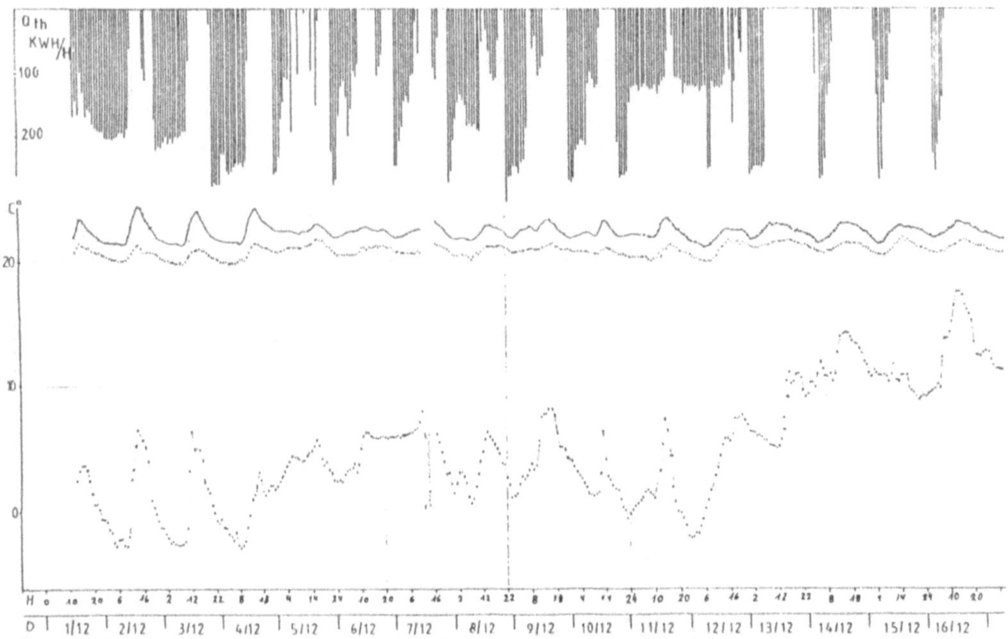

Diagram n° 2 : Overage hourly temperatures and calorific production during full scale operation period

a <u>tests and setting on period</u>

Il must be remembered that during this period, PAC 1 power has been limited to 25 % for safety reasons. This limit has been held until every parameter of the installation has been set out to satisfaction.

One effect of this limitation is that, during the abnormal cold spell of late november , some coldness was experienced in some rooms, coldness which was also due to a lack of air-tightness of the building envelop, defect mended afterwards.

On the other hand, some failures occurred during that period. Note-worthy, a PAC 1 failure due to a security switch, while unattended.

A special program sentence has been provided then to avoid the occurrence of such a failure.

Anyhow, these events did not hamper much the comfort level, even in the much exposed out - of - ground floors, under the pyramidal roof , where ambient temperature above 20° C was kept during 94 % of the time with an absolute minimum of 18,8° C.

b Full power and controlled period

Occurence of exceptional cold weather at the beginning and a milder spell at the end of the period, allowed us to ascertain every answer of the system to any weather possibilities of a normal winter season.

Mean temperatures are about set points in any circumstances. They fluctuate between 20 and 21° C for premisses average and 21-23° C in the out - of - ground floors of the pyramid where direct solar gains occasionnally raise them to 24° c.

It must be noticed concerning these overheating that the lowering of the diurnal set points has not been actuated for these rooms, on account the fear of the owner that the coldness experienced during the preceding period can come again.

IV DISCUSSION

On the basis of the data gathered, it was then possible to compute and draw in diagram n°3, the daily heat production and the unitary calorific powers of PAC 1. It is clear that internal and solar heat gains have an noticeable influence on the latest.

The highest unitary calorific demand is about 8 kW/K with an overage of 5,4 kW/K. Taking into account the 1967 degree-days 15/15, published by the official meteorological station in Uccle, it can be computed that the normal yearly heat load will be about 255 MWh or 7,4 kWh/m3 of premises

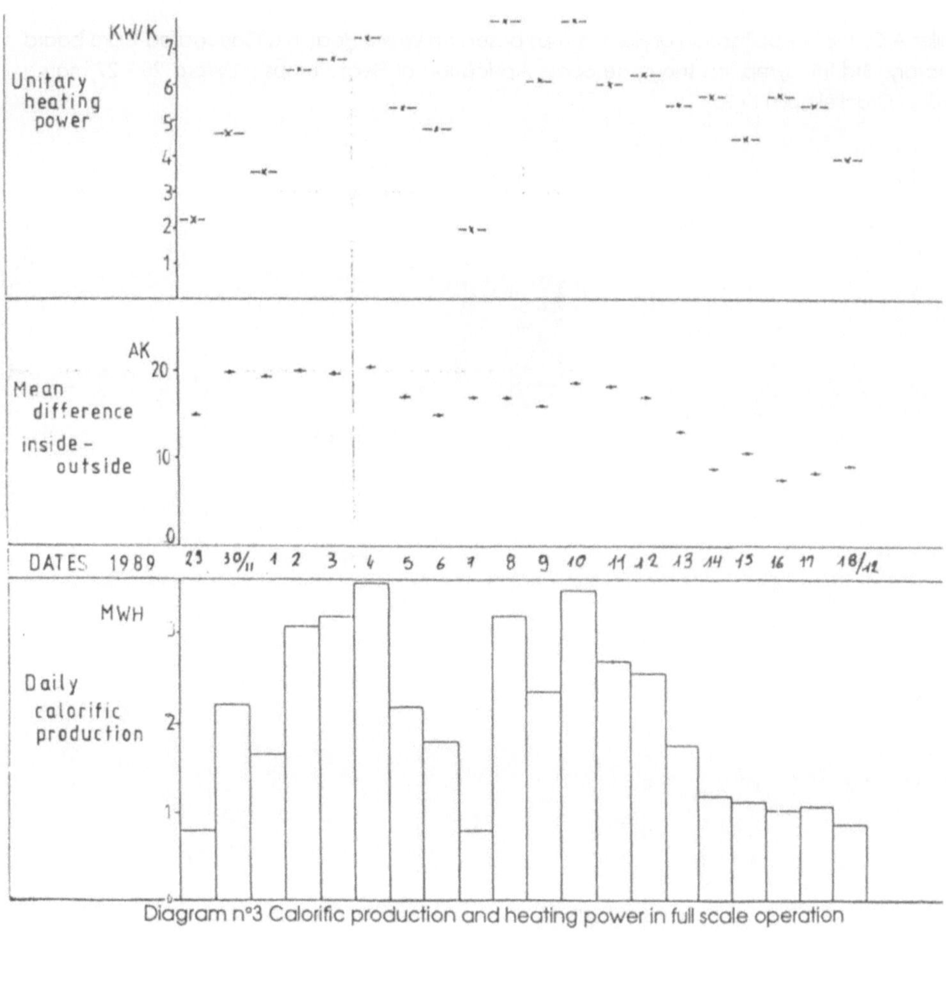

Diagram n°3 Calorific production and heating power in full scale operation

On the basis of 10 BEF/l light fuel (IPC 10 kWh) and taking into account electricity needed for auxiliary devices (0,4 BEF) annual maintenance (1,40 BEF/l) and an overal yield of 65 % (chimney, distribution and regulation) the cost price of the heat delivered by conventionnal central heating. can be evaluated to 1,82 BEF/kWh
255 MWh of boiler heat will cost 464 000 BEF annually.
On the other hand, heat pumps as designed here only use electricity, much of it at offpeak tariff, need little or no maintenance, last longer than boilers and have an overall annual coefficient of exploitation of 2.6. At present tariff, (peak 4,45 BEF/kWh, off peak 2,02 BEF/kWh) average price of electricity is about 3,4 BEF/kWh.
255 MWh of heat pump heat will cost about 333 000 BEF annually.
The 25 % of annual saving justifies the extracost of heat pumps versus boilers on a ten years payback time.

V CONCLUSION

Well designed heat pump can meet the heating needs of a well designed bioclimatic building technically as well as economically.
From previaus experience as will as the present one, with ever rising fossil fuel prices and increassing environnemental concern, it is the solution for the future in any low temperature application, specially domestic heating, which represents about one third of all energy consumption in Europe.

BIBLIOGRAPHY

Gillet A.C. 1985 - Avenir de la Pompe à chaleur dans la climatisation des bâtiments - Communication au Carrefour des Technologies - Liège - 6 au 11 novembre.

Gillet A.C. 1987 - Appllication of Heat Pumps based on Waste Heat in a Corrugated Card board Factory. 3rd Int. Symp. on the large Scale Apllication of Heat Pumps - Oxford 25 - 27 march BHRA - Cranfield pp11 - 20.

Chapter 14

AN OPEN CYCLE ABSORPTION HEAT PUMP: A SYSTEM FOR DRYING AGRICULTURAL PRODUCTS

A Bouguelia (USTHB, Algeria), H Desmorieux (ENSIAAC, Cameroon), J Labidi and P Le Goff (Laboratorie des Sciences du Genié Chemie, France)

Classical techniques for drying agricultural products have several disadvantages ; we suggest an open cycle absorption heat pump system which couples a heat pump and a drier in order to improve the drier performance.

In this paper, we present a study of two types of binary working pairs : water-lithium bromide and water-silica gel.

The principal of the process consists of circulating air in a closed cycle drier in which the agricultural product is placed. After absorption or adsorption of the moisture, the air is sent to the absorber or adsorber of the heat pump. There the air is dehumidified and reheated in order to be recycled to the drier. The hygroscopic solid or solution is regenerated with warm and dry wind, heated eventually by solar energy ; another possibility is the use of an industrial evaporator utilizing the energy of combustion of agricultural wastes.

The communication will present experimental results on a small-scale unit, and modeling the heat and mass transfer between humid air and the hygroscopic material.

1. PRESENTATION OF THE PROBLEM

Drying of foodstuffs by **direct solar radiation** has been the object of numerous technical research projets in many African countries and also in France. Although we will not discuss this subject here, we note that the principle disadvantage of the process is the low energy density of solar radiation (of the order of 4 kWh par m^2 per day) which necessitates the use of large surfaces for the drying materials and long treatment times.

A priori, it would appear very attractive to employ the drying power of hot wind with a low humidity content. The equivalent drying power of these winds is ten times larger than that of direct solar radiation. Indeed, it is normally of the order of **kilowatts** per m^2, even in the absence of the sun. However, the frequency of these winds is less predictable than that of solar radiation, and furthermore, they are often loaded with dust which could polute the materials being dried.

To overcome these two disadvantages, one method would be to use an **indirect** process, with the use of a hygroscopic solution, for example a solution of calcium chloride or silica gel.

The process proposed here consists of a gas phase flowing in a closed circuit through a dryer containing the material to be dried, and then through a packed column down which trickles a concentrated hygroscopic solution.

This solution, having thus absorbed the water, is then regenerated through contact with the hot and dry wind, eventually aided by solar radiation, or in an industrial type evaporator heated by combustion of agricultural wastes.

The process of **indirect** drying is obviously more complicated (thus requiring a larger capital investment) than the process of direct drying utilizing the same source of energy ; however, it has the following advantages :

Heat Pumps – Proceedings of the 4th International Conference, pp.149-160

1. The possibility of **continuous** operation (24 hours/day) through the intermediate storage of the solution which is only regenerated during periods where the sun and/or the wind are utilisable. This is important in the case where large quantities of perishable materials must be rapidly processed.

2. The possibility to **control** precisely the temperature and the **humidity** of the gas in a closed circuit (independently of climatic fluctuations). This is important when drying fragile materials.

3. The possibility to operate with a gas **exempt** of oxygen (for example N_2) to avoid the oxydation of fragile vegetation.

This process, which we call "DRY-ABSORB" can be used for products going from granular solids to aqueous solutions (Truchetto 1986).

In the first place, we will treat the case of dehydration of aqueous solutions : pulps, molasses, fruit-juices for which there exists a rapidly expanding international matter (Truchetto 1986). Schwartzberg (1977) has shown that there exists in the sugar and in the dairy industries a large varity of alimentary juices which could be dehydrated by hot dry air obtained by contacting a hygroscopic solution.

2. DESCRIPTION OF THE PROCESS

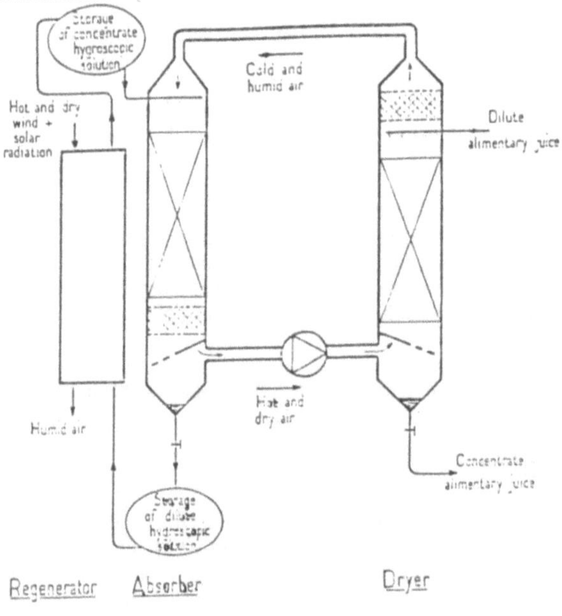

Figure 1. Process schemes using a hygroscopic solution

21. System Using a Hygroscopic Solution

The principal components of the process are two packed columns. The column on the right is irrigated with diluted alimentary juice. The latter is then contacted in counter current by the hot dry air. The cold and humid air leaving from the top of the column passes through the column on the left where it is washed co-currently by the absorbent hygroscopic solution. This exothermic absorption raises the temperature of the air while drying it. The hot dry air is then entirely recycled to the column on the right. The air passes through a demister of conventional dimensions before leaving each column. The regeneration of the absorbent solution obtained by the "dehydrating power" of the hot dry wind could be by one of the following four techniques :

1. By flowing the brine across a **quasi-horizontal black roof**. This technique has been utilized by the Russians in a similar process for air conditioning buildings (Baum and Kakabayey).

2. By flowing the brine through a solar collector containing openings to permit contacting between the liquid film and the air. This type of regeneration has been studied by Johansen and Grossman (1983) for air-conditioning processes utilizing solar energy.

3. By trickling the brine **through a conventional packed column**. The wind can then be introduced into the column by using a weather-cock.

4. By flowing the brine **across the surface of a battery of vertical streched canvasses,** positioned in advance in the direction of the dominant winds. The brine is maintained by capillarity inside the canvas and slowly flows across the surface.

The first technique has the advantage of being easy to put into use. It requires no special equipment. In addition, it has the advantage of utilizing solar radiation to heat the liquid in combination with the drying effect of the hot and dry wind.

On the other hand, it requires very large surface areas even with the anticipated dehydrating powers which are on the order of 2 to 10 kilowatts per m^2 varying according to the velocity of the wind and its relative humidity.

The third and fourth techniques are more "concentrated" in the sense that they occupy a much smaller surface area for the same amount of contact surface between the brine and the flowing air. On the other hand, they are more complicated, thus more fragile than the first technique.

22. System Using Silica Gel

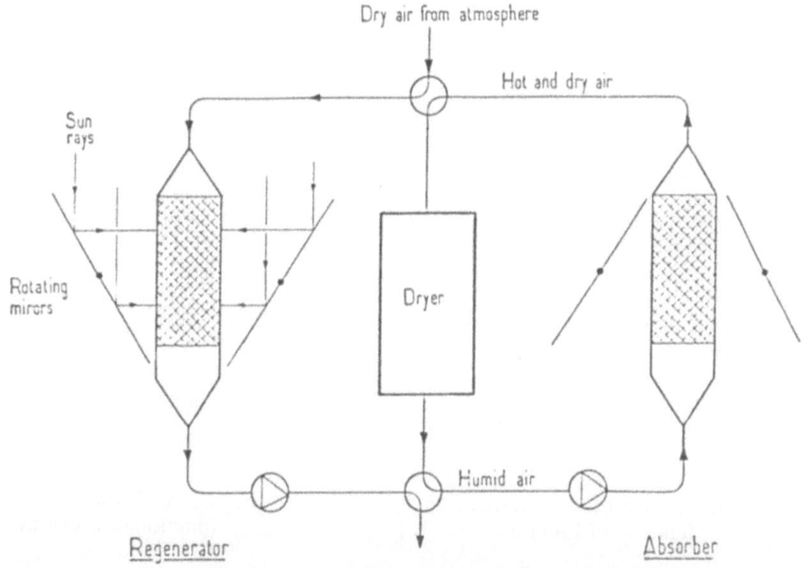

Figure 2. Process schemes using silica gel.

The use of a fixed bed of solid adsorbant in the place of a hygroscopic solution simplifies the process but imposes the use of discontinuous operation. Nevertheless the simultaneous use of two fixed beds working alternately allows a pseudo-continuous operation. The adsorbant can be regenerated in a second reactor, identical to the first, by means of direct solar radiation and by external air penetrating in the bed. The porous medium is retained between two aluminium sheets with blackened outside faces. Moveable mirrors catch the suns rays and direct them to one or the other of the two reactors according to which is in the desorption phase or the adsorption phase. Figure 2 gives a schematic diagram of an arrangment for such a pseudo-continuous operating with the two systems of regeneration. The two columns operate in alternance as an adsorber and as a regenerator. The direction of flow is reversed at the end of the cycle so as to push the concentration fronts back up the columns.

3. EXPERIMENTAL APPARATUS

31. Experimental Apparatus Using a Hygroscopic Solution
The experimental apparatus used for the hygroscopic solution is shown in Fig. 1. The columns have a diameter of 10 cm and length 2.55 m. We have tested two packing materials :

- standard 10 mm porcelain Raschig rings with a porosity of 68 % ;
- type BX Sulzer drums (Sulzer 1970) with an ordered structure of high porosity (96 %) which permits large air flow rates with a small pressure drop.

32. Experimental Apparatus Using a Fixed Bed of Silica Gel

321. Investigation of the adsorption
The experimental equipment is essentially a rectangular cross-section fixed bed (2 cm x 20 cm) with a height of up to 80 cm. The silica gel is held in place between two plates which can attain a temperature of 120 °C. The humidity of the inlet air is controlled by a valve which allows mixing dry air with the humid air from a bubble column. The relative humidity of the inlet air can vary between 30 to 80 %. In the adsorption, the air temperature is 21 °C. The silica gel is regenerated by air at 27 °C on a 25 % relative humidity. The retaining plates are kept at 80 °C. The air flow rate is measured with a rotameter and can vary from 0.78 to 3.10 m/sec.

322. Investigation of the regeneration
Tests of regeneration have been carried out in a tropical region at N'Gaoundéré in Cameroon. Saturated silica gel was placed in a 10 cm diameter column. Outside ambiant air was passed through the column by means of a centrifugal fan. The relative humidity and the temperature of the air were measured at the fan inlet and at the column outlet. The temperature was also measured at the inlet to the column.

4. EXPERIMENTAL RESULTS

41. Apparatus Using a Hygroscopic Solution

We have measured the temperature and relative humidity of the air at the entrance and exit of each column for various liquid flow rates at a **constant gas flow rate** fixed at approximately 80 % of the maximum. The results of these measurements are shown in Table I and Figs. 3a and 3b.

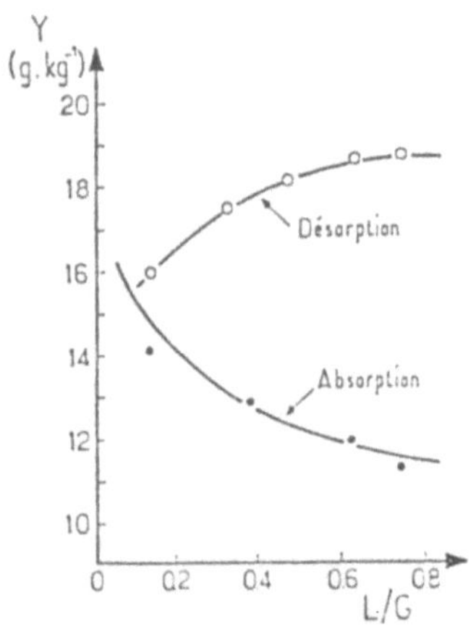

Figure 3a. Absolute humidity as a function of L/G ratio.

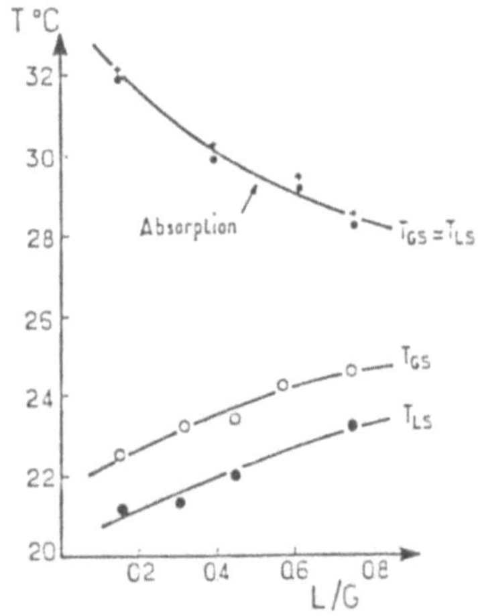

Figure 3b. Temperature as a function of L/G ratio

		ABSORBER (dehumidification of air)				DESORBER (humidification of air)				
liquid and air flow rate	G (kg h^{-1})	80	80	80	80	80	80	80	80	80
	L (kg h^{-1})	10	30	50	60	12	25	35	45	60
	L/G	0.13	0.38	0.63	0.75	0.15	0.31	0.44	0.56	0.75
air in	T_{GE} (°C)	24	24	24	24	29.2	29.2	29.2	29.2	29.2
	HR_E (%)	98	98	98	98	47.3	47.3	47.3	47.3	47.3
	T_E (g kg^{-1})	18.5	18.5	18.5	18.5	12	12	12	12	12
air out	T_{GS} (°C)	32	30	29.2	28.3	22.5	23.2	23.4	24.2	24.5
	HR_S (%)	46.6	48.2	47.3	46.6	92.8	96.4	99	96.9	95.7
	Y_S (g kg^{-1})	13.9	12.8	12	11.2	15.9	17.3	18	18.5	18.6
liquid in	T_{LE} (°C)	22	22	22	22	27	27	27	27	27
	x_E (%)	40	40	40	40					
liquid out	T_{LS} (°C)	32	30.4	29.5	28.5	21.2	21.4	22	22.1	23.2
	x_S (%)	36.7	38.3	39	38.9					

Table I. Experimental data.

It can be seen that at L/G ratios greater than 0.6, the humidity at the exit remains approximately constant for both absorption and desorption. We will use this value as the optimum operating point for the installation.

The increase in mass and heat transfer rate with increasing liquid flow rate can be explained by the fact that dehumidification and humidification of air both tend toward isothermal processes. Indeed, for absorption for example, the heat produced is carried away by the liquid whose flow rate becomes larger and larger.

42. Apparatus Based on Silica Gel

421. Adsorption

Experiments have been performed with a rectangular cross-section fixed bed to determine the influence of various parameters on the adsorption of water vapour. These were : the flow rate of gas through the bed, the humidity at the bed inlet and the depth of the bed. All other parameters were kept constant. The measured breakthrough curves are presented in Figs. 4, 6 and 8. Figures 5 and 7 show the variation of the mean flux of water adsorbed. This is defined as $m_e = m_e/t_f$ were m_e is the cumulative amount of water adsorbed given by the surface between the breakthrough curve and the straight line $x/x_0 = 1$. It can be seen that the mean flux of water adsorbed m_e increases linearly with the flow rate and with the inlet humidity. Figure 8 shows that an increase in the bed depth results in a reduction in the humidity of the air at the outlet. This may be attributed to the increase in the contact time with the increase in the bed depth.

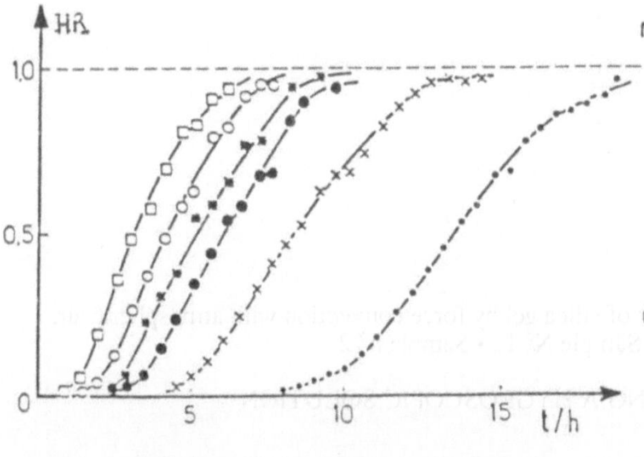

Figure 4. Relative humidity at the outlet, vs time (in hours) for various interstitial flow rates

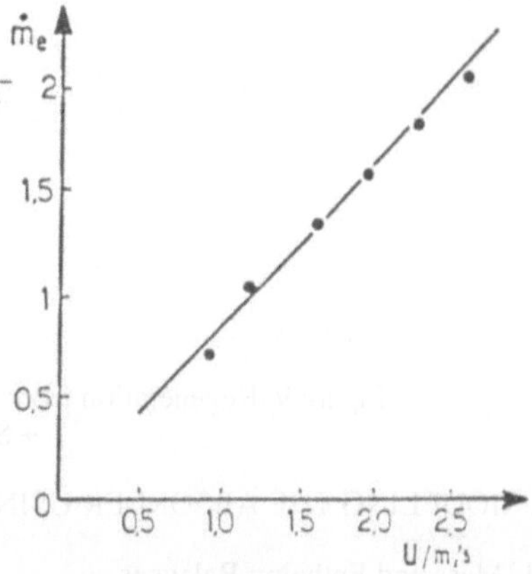

Figure 5. Mean flux of water transfered as a function of the interstitial flow rate.

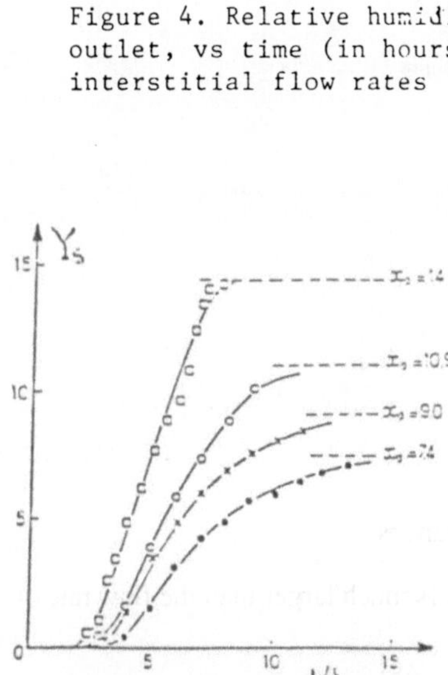

Figure 6. Absolute humidity (in g/kg) at the outlet, vs time (in hours) for various absolute humidity at the inlet

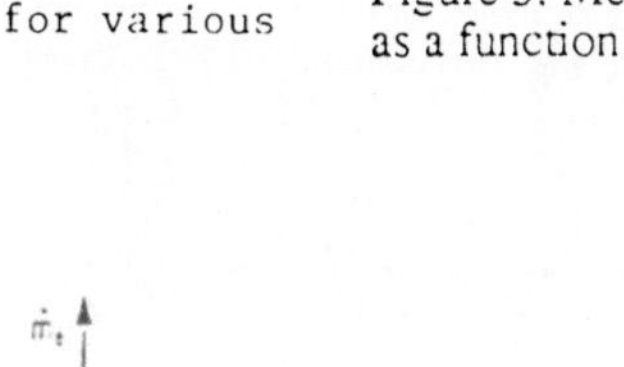

Figure 7. Mean flux of water transfered as a function of the humidity at the inlet.

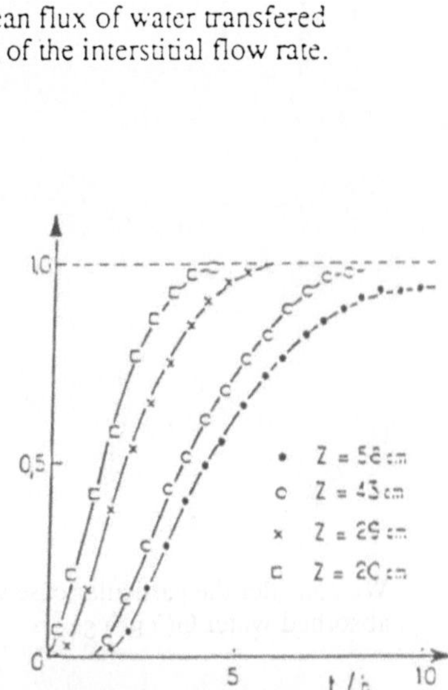

Figure 8. Relative humidity at the outlet, vs time (in hours) for various be depths

422. Regeneration
We have carried out experiments of regenerating silica gel in a tropical region using two samples : one of 685 g and the other of 800 g of gel (see Fig. 9). The temperature at the inlet to the column may be considered to remain constant over the duration of the experiment (90 minutes). In experiment n° 1, it was 37 ± 1 °C and in experiment n° 2, it was 34 ± 1 °C. The slopes of these curves at the beginning of desorption can be attributed to the evaporation of free water present in the packed bed. The fronts obtained demonstrate that the silica gel can be regenerated at this relatively low temperature and low humidity. At the end of the experiment, there is only a very small difference from equilibrium. According to the equilibrium curve (Nanakasse 1986) after regeneration, the silica gel in equilibrium with the air at 13 % and 25 °C, is 7 g of water per 100 g of gel. Pre-heating the air by solar heating would increase the efficiency of regeneration.

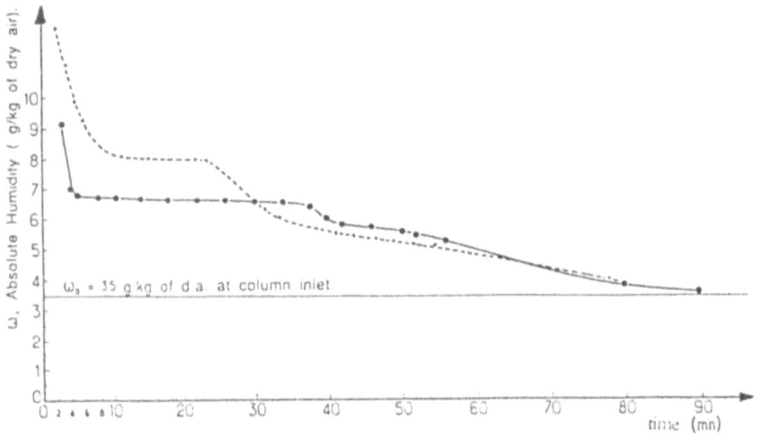

Figure 9. Regeneration front of silica gel by force convection with atmospheric air.
+ Sample N° 1 ; • Sample n° 2.

5. MODELING THE ABSORBER USING A HYGROSCOPIC SOLUTION

51. Mass and Enthalpy Balances

Overall water and enthalpy balances for the absorption column can be written as (see Fig. 10)

$$L_E(1-x_E) + GY_E = L_S(1-x_S) + GY_S$$

$$L_E h_{LE} + G h_{GE} = L_S h_{LS} + G h_{GS}$$

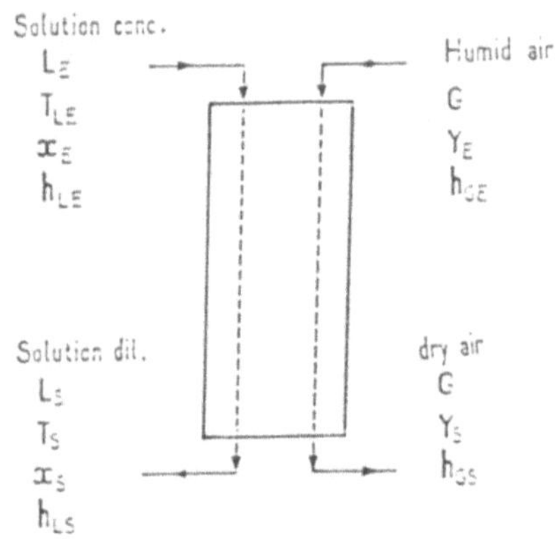

Figure 10. Mass and enthalpy balances.

We consider the particular case where the total solution flow rate L_E is much larger than the flow rate of absorbed water $G(Y_E-Y_S)$.

The liquid flow rate L can be taken as being constant in the enthalpy balance. By assuming that the specific heat of the solution remains constant, the enthalpy balance becomes

$$\frac{h_{GE}-h_{GS}}{T_{LE}-T_{LS}} = \frac{-L\ \overline{C_{PL}}}{G} = \text{constant}$$

This equation describes the operating line or balance line, with constant slope, that is to say the trajectory of the point whose coordinates are the gas enthalpy (h_G) and the liquid temperature T_L (Fig. 11).

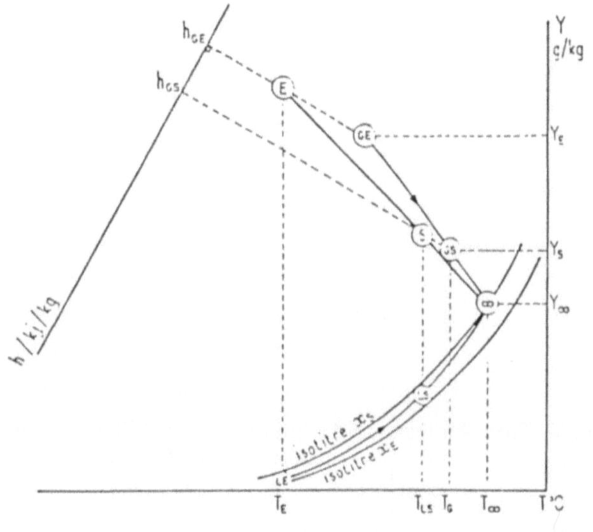

Figure 11. Enthalpy-temperature-humidity diagram (Veron-Casari).

The points on the operating line, representing the entrance and exit of the absorber, are respectively E and S.

The composition of the gas phase at the inlet is represented by the point G_E with coordinates (h_{GE}, Y_E, T_{GE}) and the composition of the liquid phase at the entrance is given by the point L_E at temperature T_{LE} and mass fraction x_E*.

Likewise, the points (GS) and (LS) represent the outlet from the absorber.

The point (∞) represents the final state which can be attained at infinite contact time, the two phases thus leaving in full thermodynamic equilibrium.

52. Limiting Case of a Perfectly Efficient Absorber
Enthalpy variations in the gas phase are given by

$$(h_{GE}-h_{GS}) \approx Cp_a(T_{GE}-T_{GS}) + L_V(Y_E-Y_S) + Cp_V(Y_E T_{GE}-Y_S T_{GS})$$

Knowing that the last term in this expression is negligible and combining it with the preceding expression for the point (∞) i.e.,

$$T_{LS} = T_{GS} = T_\infty$$

we obtain

$$Y = Y_E + \frac{Cp_a(T_{GE}-T_\infty)}{L_V} + L\ \overline{Cp_L}\ \frac{(T_{LE}-T_\infty)}{G}$$

This expression shows that the absolute humidity of the exiting air Y_∞ is a **decreasing linear** function of the temperature T_∞, **independent** of the **absorbent** properties of the hygroscopic solution.

The coordinates of the limit point are shown in Fig. 12 respectively for a 40 % $CaCl_2$ solution and a 55 % LiBr solution at the following conditions

$T_{GE} = 24\ °C$ $\quad$ $Y_E = 18.5\ g\ kg^{-1}$

$T_{LE} = 22\ °C$ $\quad$ $L/G = 0.60$

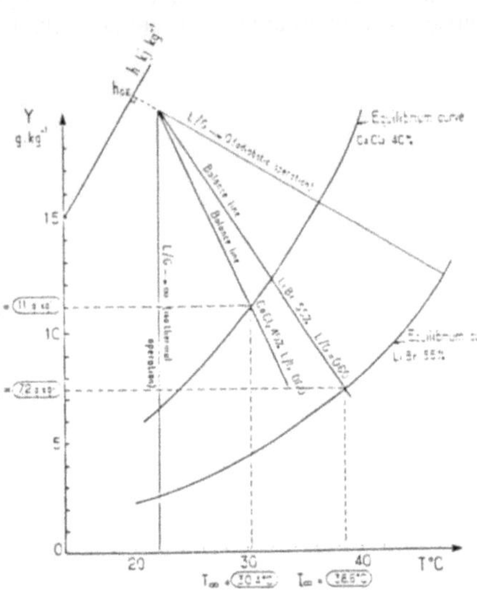

Figure 12. The coordinates of the limit-point for 40 % $CaCl_2$ solution and 55 % LiBr solution in each case of absorption.

In this figure, the influence of the L/G ratio for values between 0 (quasi-adiabatic process) and infinity (quasi-isothermal process) can be seen.

Thermal coupling between the two columns (dryer and absorber) permits isothermal operation of the system. Using the calcium chloride solution as an example, it can be seen from Fig. 12 that the air leaving the absorber is approximately 34 % dryer when thermal coupling is employed. There exist several techniques for realizing thermal coupling :

1. Schwartzberg (1977) has proposed a vertical tube exchanger.
2. A plate exchanger can also be employed by passing the hygroscopic solution on one side and the alimentary juice to be concentrated on the other.
3. An exchanger could be integrated into the column ; however, the design of such an apparatus would be more difficult.

In Fig. 13, we have presented, as a function of L/G ratio, some of the experimental data for Y_S and T_S, and the curve of limiting values, Y_∞ and T_∞, which would be obtained for perfectly efficient absorber. The efficiency of the absorber is then given by the relation

$$\eta_L = \frac{T_{LS}-T_{LE}}{T_\infty - T_{LE}} \qquad \text{and} \qquad \eta_G = \frac{Y_E - Y_S}{T_E - T_\infty}$$

We thus obtain, for OL/G = 0.60 for the two cases : $\eta_L \approx 89\ \%$ and $\eta_G \approx 88\ \%$.

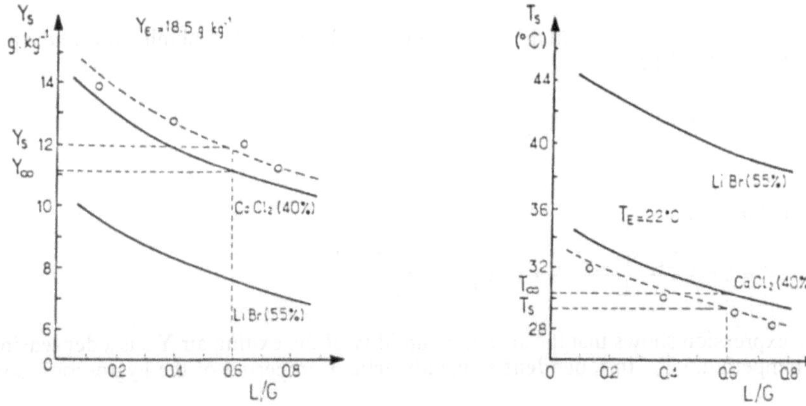

Figure 13. Comparison of the experimental results with the theory.

53. Absorption Kinetics
It is well known (Brown 1950) that for most humidification/drying operations of low humidity air, the transfer resistance is principally localized in the gas phase. In other words, the interfacial temperature T_i is only slightly higher than that in the bulk of the liquid phase T_L, and thus one may use either of these values.

It can be shown (Foust 1980) that the enthalpy of the gas phase plays the role of an **intensive** quantity. In other words, the flux transfer density is proportional to the enthalpy difference between the bulk of the gas phase and the interface

$$d\phi = k_G(h_G - h_{Gi})dA$$

where $dA = a\Omega dZ$

Integrating this leads to the well known relation

$$k_Ga = \frac{G}{z\Omega}\int_{h_{GE}}^{h_{GS}} \frac{dh_G}{(h_G - h_{Gi})}$$

Figure 14 shows the variation of the mass transfer conductance as a function of the liquid flow rate calculated from the above integral for the Sulzer BX packing and a calcium chloride solution.

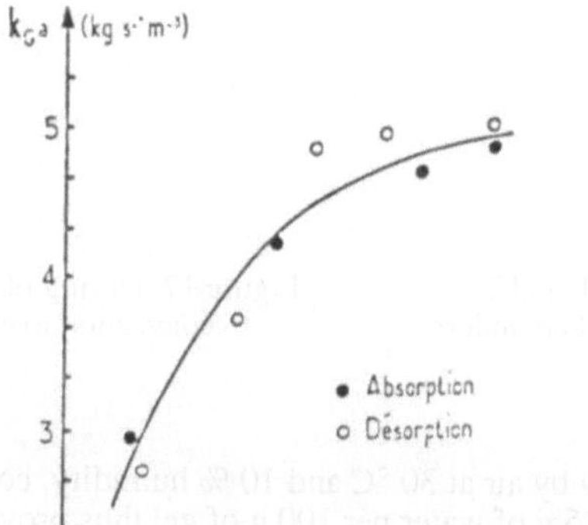

Figure 14. Coefficient of mass transfer as a function of the liquid specific rate.

6. MODELLING THE FIXED BED

The basic equation for medelling adsorption in a fixed bed is the differential mass over a thin slice of the bed between Z and Z+dz. This equation is mathematically complex and does not have an analytical solution if mass transfer resistances and hydrodynamic dispersion are included. For the case of isothermal adsorption in a fixed bed, all the models available in the literature give solutions valid for different limiting cases. Under the experimental conditions used here the best agreement is obtained with the Thomas model. This model accounts for certain thermodynamic (adsorption isotherm), and kinetic factors (rate of adsorption) (Fig. 15). Details of this modelling is to be published elsewhere.

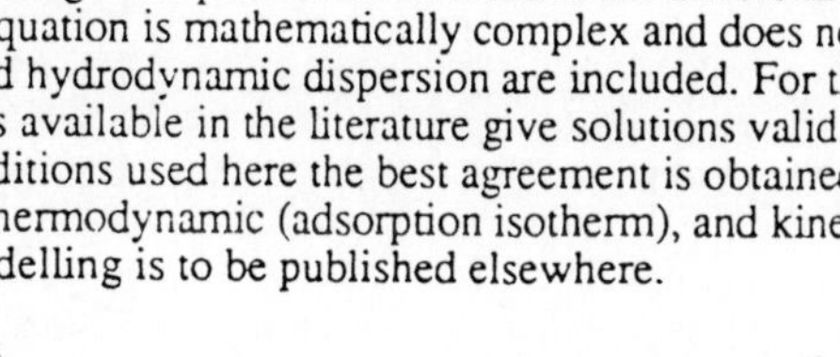

Figure 15. Comparison of an experimental breakthrough curve with predictions from a theoretical model.

7. AN APPLICATION OF THE PROCESS IN A TROPICAL REGION

The results obtained in both the adsorption and regeneration phases indicate that the "Sechadsorb" process could be applied in tropical regions. In such regions, the temperature and humidity vary a great deal between the night and the day, especially between the months of November to May. During this period, food products can easily be dired during the day by means of solar energy and/or dry air. At night when the humidity of the air prevents or slows down any drying action the products may re-humidify and their quality may be affected. Figure 16 based on three hourly mean meteorological data for February (Jannot 1988) and shows that the variation during the day of ΔP the difference between the saturated vapour pressure and the partial pressure of water vapour gives a sufficient motive force for evaporating water. This is confirmed by Fig. 17 which shows the drying curve of bananas obtained by forced convection of ambiant air (February in N'Gaoundéré). At night the weight loss is either very small or null. To remedy this it would be useful to be able to store the energy available during the day time and use it at night for de-humidifying the air circulating in closed circuit in the dryer. The time required for drying a product would then be considerably reduced.

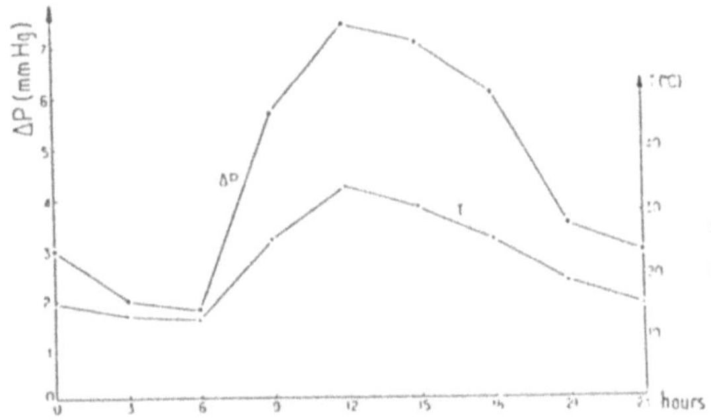

Figure 16. Daily evolution of ΔP and T during the month of February in N'Gaoundéré.

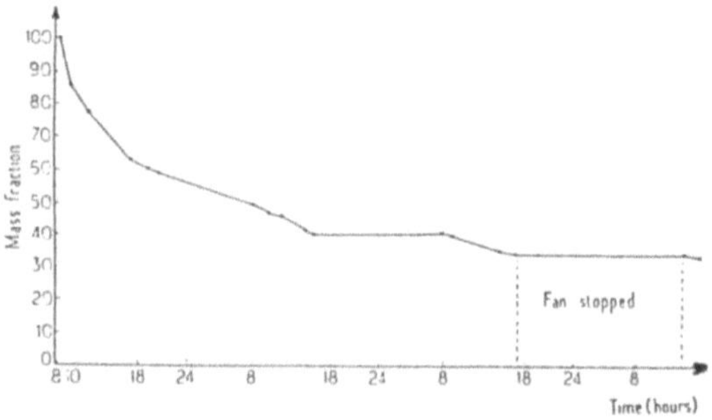

Figure 17. Drying of banana by force convection to external air.

Silica gel, regenerated during the day by air at 30 °C and 10 % humidity, contains less than 10 % water and during the night could absorb 20 to 25 g of water per 100 g of gel thus providing air with a relative humidity of less than 60 %. Therefore at night the air is de-humidified before entering the dryer thus at least maintaining the day time drying efficiency. After passing through the dryer the air is then rejected to atmosphere, though a closed circuit system could be used.

As shown in Fig. 18, this apparatus required for this drying process could be of relatively rustic construction based on locally available materials. Regeneration being carried out during the day at the same time as the drying of products laid on trays. A low cost solar cell made of black painted galvanised steel sheet (efficiency 30 %) heats air before drying and before de-humidifying, thus increasing the efficiency of the system.

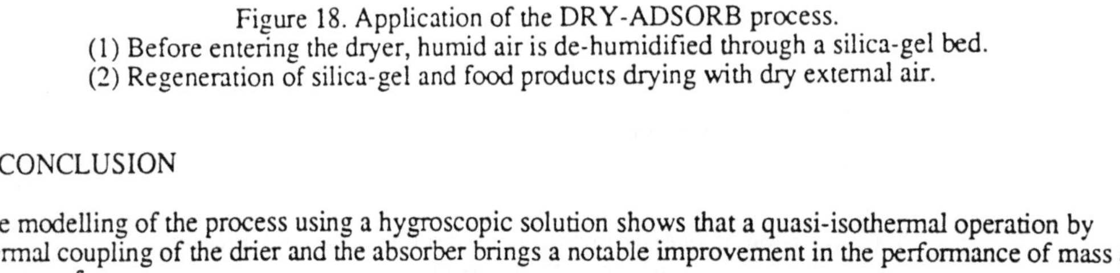

Figure 18. Application of the DRY-ADSORB process.
(1) Before entering the dryer, humid air is de-humidified through a silica-gel bed.
(2) Regeneration of silica-gel and food products drying with dry external air.

8. CONCLUSION

The modelling of the process using a hygroscopic solution shows that a quasi-isothermal operation by thermal coupling of the drier and the absorber brings a notable improvement in the performance of mass and heat transfer.

The process based on the use of silica gel is simpler and therefore cheaper than the system based on the hygroscopic solution. However modelling the pseudo-continuous operation is a complex problem. Thermal coupling between the drier and the absorber would also give better efficiency but would make the system much more complicated.

NOMENCLATURE

a	specific surface
Cp	specific heat
G	flow rate of dry air
h	specific enthalpy
k	mass transfer coefficient
L	liquid flow rate
T	temperature
Y or x	absolute humidity (equilibrium)
Z	height of packing or bed
t_f	end time
HR	relative humidity

Subscripts

a	air
E_o	inlet
G	gas
i	interface
L	liquid
s	outlet
v	vapour
∞	conditions at infinite time

9. REFERENCES

BAUM, Y.A. and KAKABAYEY, A., The use of solar radiation for creating confortable temperature conditions in Turkmenian SSR, The Phys. and Tech. Inst. of the I.S.S.R. Academy of Sci., URSS.

BROWN, G.G., and col., Unit Operations, John Wiley & Sons, New-York Chapman & Hall Limited London, pp. 242-258.

FOUST, and col., (1980), Simultaneous heat and mass transfer humidification, Principles of unit operations, 2nd Edition, pp. 420-440.

JANNOT, Y., (1988), Traitements des données météorologiques du Nord Cameroun, Communication interne.

JOHANNSEN, A., and GROSSMANN, G., (1983), Performance simulation of regenerating type solar collector, Vol. 30, n° 2, Solar Energy, pp. 87-92.

LE GOFF, P., (1980), Coordonnateur "Energétique Industrielle" en trois tomes, Technique et Documentation, Paris.

SCHWARTZBERG, H.G., (1977), Energy requirements for liquid food concentration, Food Technology, pp. 67-76.

SCHWARTZBERG, H.G., and col., (1977), Food evaporators driven by solar concentrated brines, Food Technology.

SHERWOOD, T.K., and col., (1975), Design of fixed-bed sorption and ion exchange devices, Mc Graw-Hill Chemical Eng. Series.

SULZER, (1970), Colonnes de séparation pour distillation et absorption. Revue Technique, n° 3.

TRUCHETTO, (1986), Le marché international des pulpes et concentrés de fruits tropicaux : les perspectives de pénétration des produits Camerounais, Rapport ENSIAAC, Cameroun.

Chapter 15

HEAT PUMPS WITH DIFFERENT ENERGY COLLECTORS IN ARMY BARRACKS (VOLKACH, GERMANY): RESULTS FROM THREE HEATING SEASONS (1986-1989)

J Ficke, U Heinemann and E Hümmer (Physikalisches Institut der Universität Wurzburg, Germany)

ABSTRACT

A combination of conventional gas/oil burners and heat pumps driven by gas engines supplies heating energy and hot water for army barracks in Volkach/ W.Germany. Seven different anergy collectors are analyzed: two energy roofs, one energy facade, three energy staples and one evaporator with forced ventilation. Main objective is to study these heat pump systems (HPS) under comparable meteorological conditions and similar user habits. Continuous monitoring of about 750 sensors is provided. Experiences and results of three heating seasons (1986-1989) are presented. The heat pumps covered 22% (winter 1986/87), 36% (1987/88) and 38% (1988/89) of the heating demand, respectively. This was less than could be anticipated (36%, 52% and 59%, respectively) from design parameters. The observed deficit is caused by malfunctions of the HPS and by problems with the control units. The coefficients of performance (cop) varies from 1.05 to 1.56. The primary energy saved by the HPS with respect to a conventional heating system was 11%, 16% and 16% in these three heating seasons, versus values of 16%, 24% and 27%, respectively, expected for a cop of 1.45 and regular HPS operation. The scientific research program is supported by the BMFT (German Federal Ministry for Research and Technology).

1. ENERGY CONCEPT AND AIM OF THE INVESTIGATIONS

Generally army barracks are equipped with one large conventional heating system. The Mainfranken Barracks at Volkach, however, are supplied with nine independent heating units (Fig. 1), each of which meets the demand of one to six buildings. In seven of these units heat pumps driven by gas engines are used in addition to conventional gas/oil burners. Different types of anergy collectors extract heat from surrounding air, some also use solar radiation. The heat pumps themselves are comparable in design.

All heating systems had been installed until June 1986. Continuous data collection from 750 sensors in 15s-intervals started November 1987. Data collection and evaluation is performed by our institute.

Aims of the supporting scientific program are

- the continuous registration of averaged 30min-data during several heating seasons,
- evaluation of relevant energy flows,

Heat Pumps - Proceedings of the 4th International Conference, pp.161-176

- investigation of the transient behaviour of the system (based on data collected in 30s-intervals) for a detailed characterization of the components and for elimination of problems in controlling,
- optimization of the systems with respect to a minimal primary energy input,
- comparison between the different anergy collectors, and
- proposals for future heat pump conceptions.

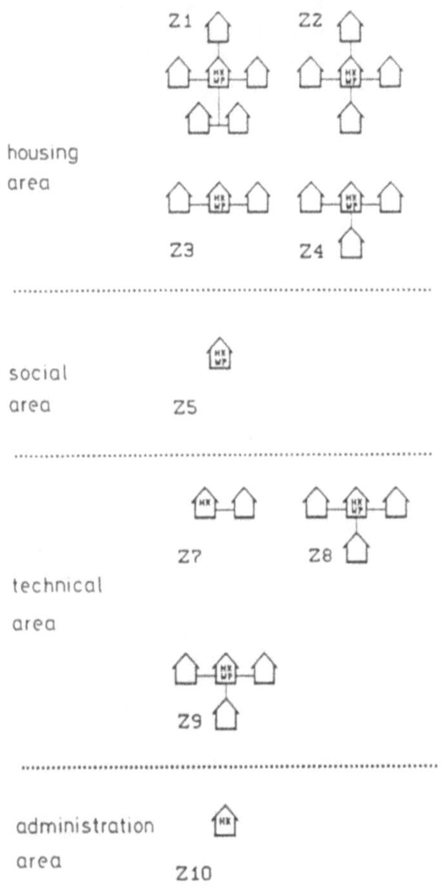

Fig. 1: Heating centers with attached buildings.

The uniqueness of this research project results from the fact, that different collector systems can be studied in practice under comparable meteorological and user conditions.

Total investments for the nine heating systems amounted to 12 million DM, of which 9 million DM are spent for the heat pump/collector installation and the data acquisition system.

2. DESCRIPTION OF THE DIFFERENT HEATING SYSTEMS

Seven of the nine heating systems are equipped with compressor heat pumps driven by gas engines. These systems function in a bivalent manner: for outdoor temperatures below +3°C gas/oil burners supply all the heat, above +3°C the heat pumps are supposed to operate. The heating systems in general are rather complex. In the technical area (garages, workshops and storages) centers Z7, Z8 and Z9 only provide heating energy. The housing area centers Z1, Z2, Z3 and Z4 in addition supply domestic hot water. Center Z5 delivers the energy for the social area (kitchen and cafeteria) and therefore is equipped with two high pressure steam vessels. Center Z10 belongs to the administration area (operating since 1989). Center Z6 was never built.

center	burner kW	heat pump kW	anergy collector	area
Z 1	570	290	energy roof absorber, copper	housing
Z 2	650	325	energy roof absorber, aluminium	housing
Z 3	425	196	energy staples, aluminium	housing
Z 4	496	229	energy staples, copper	housing
Z 5	730	400	forced ventilation evaporator	social
	920	-	-	social
Z 7	200	-	-	technical
Z 8	815	173	energy staples, plastic	technical
Z 9	896	220	facade absorber, steel	technical
Z10	400	-	-	administr.

Table 1: Characterization of the heating centers.

In Table 1 the characteristics of the different systems are listed. In centers Z1 to Z5 heat pumps are capable of delivering the total energy demand for temperatures above +3°C. In centers Z8 and Z9 the heat pumps normally are supported by the gas/oil burners up to +10°C. Two types of anergy collectors are used:

- energy staples which extract anergy from the surrounding air,
- absorber roofs or facades which may be viewed as uncovered solar collectors; maximum temperatures of 60°C for full insolation are reached in summer; in this case direct heating of the hot water tanks is possible (heat pumps off). During sunny winter days the heat pumps may be operated below +3°C.

The scheme of one of the most complex system (Z1) is shown in Fig. 2.

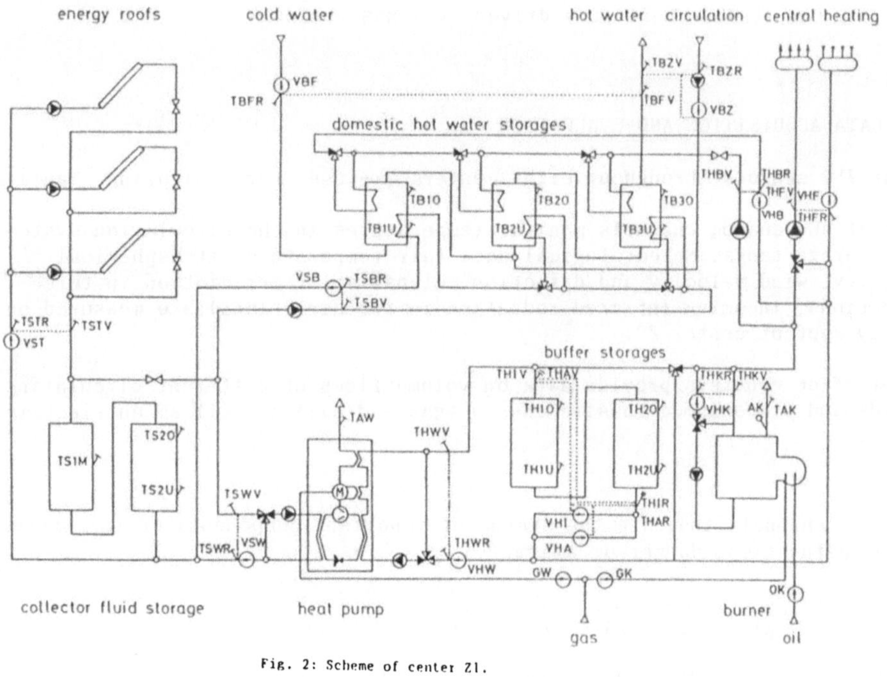

Fig. 2: Scheme of center Z1.

3. HEAT PUMPS

The heat pump aggregates consist of the gas engine, the refrigerant circuit (evaporator, compressor, condenser, expansion valve) and the heat exchangers for the utilization of energy from the engine cooling circuit and the exhaust (Fig. 3).

The aggregates are noise shielded and are mounted on a separate concrete basement for vibration isolation.

To prevent detrimental frequent switching the heat pumps are either on or off for at least 30 minutes. The variable heating is met via the use of storage tanks.

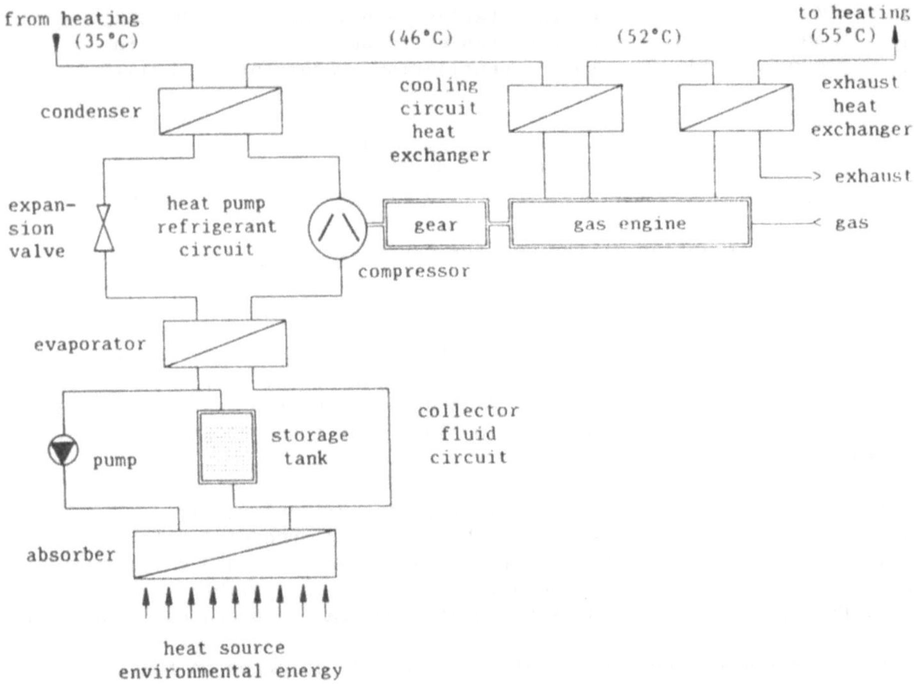

Fig. 3: Compression heat pump driven by a gas engine.

4. DATA ACQUISITION AND EVALUATION

About 750 sensors throughout eight centers are used for monitoring (Table 2).

Most of the analog channels monitor temperatures in the circulation system and the storage tanks. Meteorological data (air temperature, atmospherical humidity, wind velocity and direction, global solar irradiation in three directions, incoming infrared radiation in two directions) are measured on the energy roof of center Z2.

Inputs from counters provide data on volume flows of different circulating fluids and of consumed fossile energy (gas and oil) as well as on electrical power.

Digital channels indicate the status of important components of the system, e.g. malfunction, danger or alarm.

center	Z1	Z2	Z3	Z4	Z5	Z7	Z8	Z9	Σ
analog	41	76	37	34	46	5	19	24	282
counter	17	23	18	15	23	5	10	11	122
digital	38	36	32	36	73	29	52	58	354
total	96	135	87	85	142	39	81	93	758

Table 2: Numbers and types of data channels.

For Z1 the position of the sensors is shown in Fig. 2.

The scanning of the sensors is performed by an intelligent substation in each center. Raw data are transmitted via serial interfaces to a master station, which is connected to two independently working personal computers.

The personal computers run programs for continuous data compression and archivation, they display the actual operating status of the systems and the data trends.

5. RESULTS

For comparison of different heating seasons meteorological data of the corresponding periods have to be accounted for. Totalisation of positive temperature differences between indoor and outdoor gives the number of degree-days for each heating season (Table 3).

period	mean outdoor temperature °C	degree-days K•d
1986/87	7.8	4388
1987/88	9.6	3617
1988/89	10.1	3525

Table 3: Averaged outdoor temperatures and degree-days in Volkach/W.Germany for three heating seasons; the degree-day values are derived for indoor temperatures of 20°C, considering only periods with outdoor temperatures below 15°C.

5.1 Energy Flows

From the continuously sampled data of year 1988 the energy flows for seven heating systems are derived (Fig. 4).

Energy inputs are

gas and oil consumption of the burners	GK, OK,
gas consumption of the heat pumps	GW,
environmental energy	
extracted by the collector	QST,
and fed into the heat pump	QSW,
in Z5	QKV.

Thermal losses are depicted horizontally:

losses of the burners	LK,
losses of the heat pumps	LW,
losses of the buffer storage tanks	LH,
losses of the hot water storage tanks	LB,
losses in hot water circulation circuit	QBZ.

The utilized energies are

for heating demand	QHF,
for hot water demand	QBF.

These energies are delivered by

burners	QHK,
heat pumps	QHW,
collector fluid at high temperatures (directly used)	QSB,
high pressure steam circuit in Z5	QNB.

Additional energy flows are

buffer storage in- and outputs	QHI, QHA
energy transferred from heating fluid to domestic hot water system	QHB.

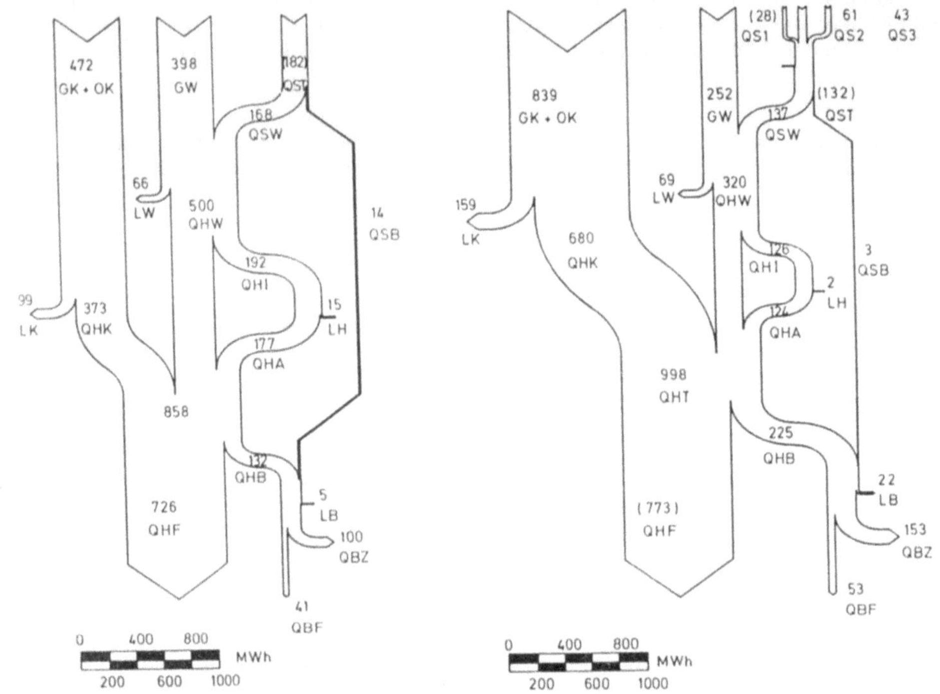

Fig. 4 a): Energy flows in MWh in 1988 for Z1 and Z2.

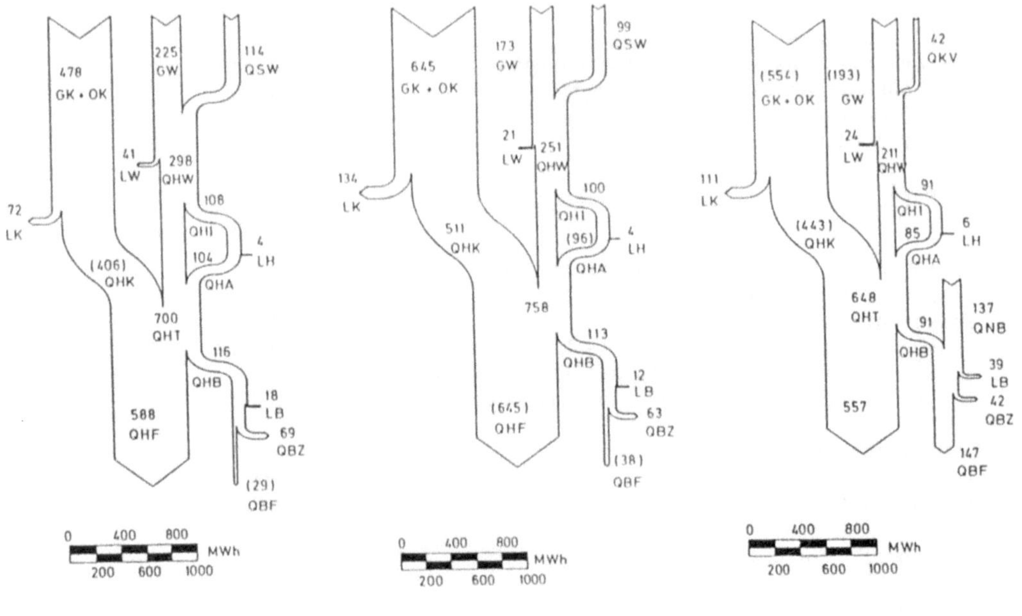

Fig. 4 b): Energy flows in MWh in 1988 for Z3, Z4 and Z5.

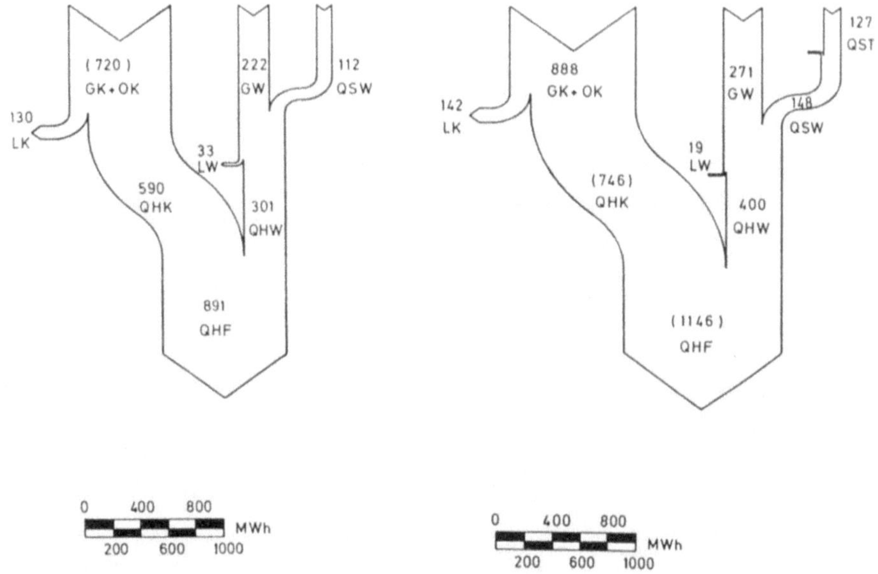

Fig. 4 c): Energy flows in MWh in 1988 for Z8 and Z9.

5.2 Contributions From Heat Pump Operation

The measured monthly energy consumption including hot water demand is shown in Fig. 5. Also shown are the actual and theoretical contributions of the heat pumps. The latter are based on measured outdoor temperatures and demand for heating energy; it is assumed that the heat pumps operate above +3°C outdoor temperature and supply the demand up to their rated power. The hatched areas indicate the differences between theoretical and actual contributions. It is obvious that during the cold months (November to March) a large demand coincides with small (theoretical and actual) contributions from the heat pumps; as can be expected from the design parameters (e.g. +3°C as switching temperature), large absolute contributions can only be found for spring and autumn. In the summer months only a small demand for domestic hot water remains.

It should be pointed out that the winters 1987/88 and 1988/89 were untypically warm, allowing larger contributions from the heat pumps.

Surprisingly large are the deficits, indicated by the hatched areas, which can be explained by frequent malfunctions of the control systems and failure of components. We want to underline, that the heat pumps reach or even exceed the rated power, if they are operating at all.

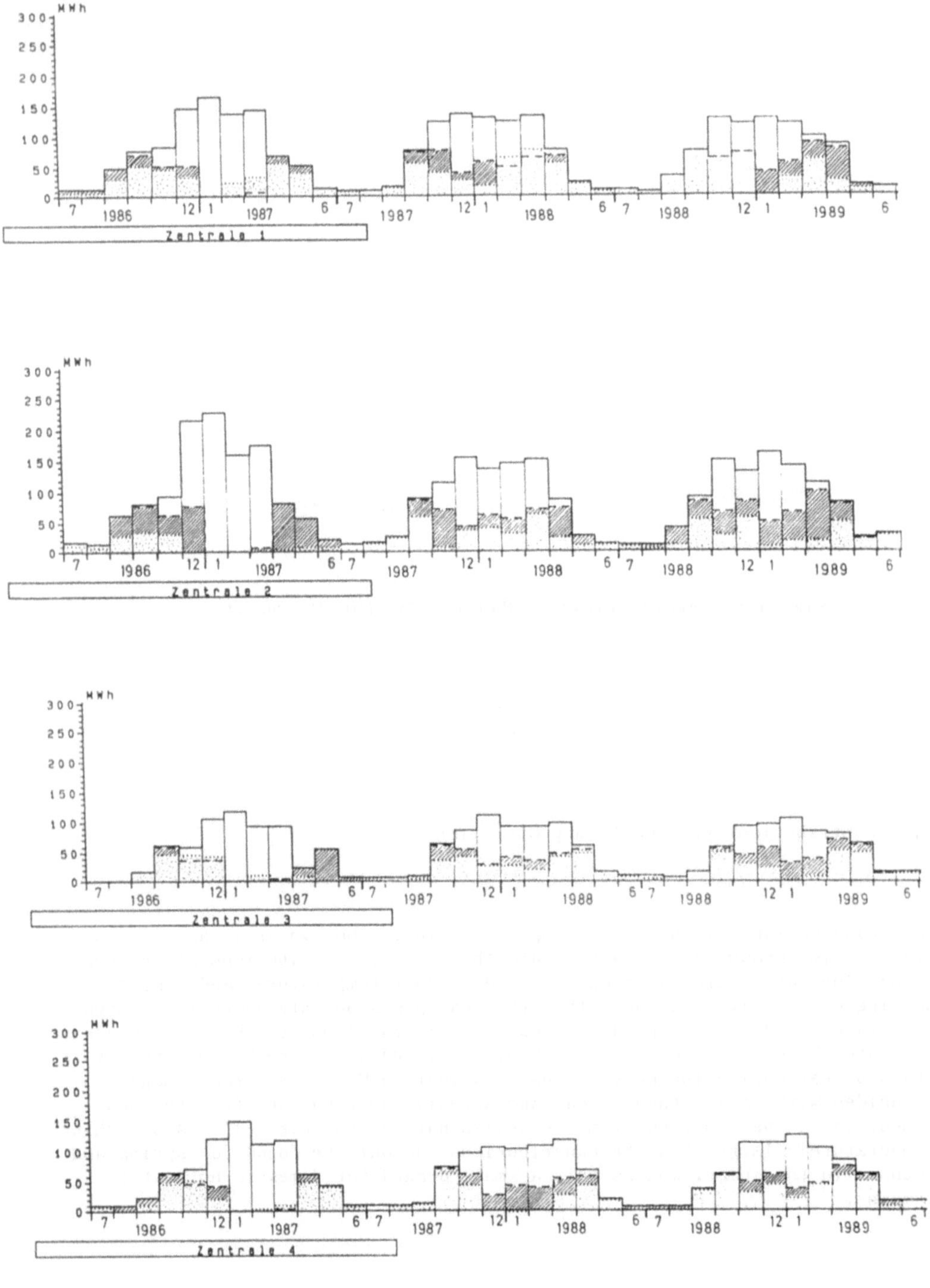

Fig. 5 a): Measured monthly energy consumption (solid lines) as well as actual (dotted lines) and theoretical (dashed lines) heat pump contributions from 1986 to 1989 for Z1, Z2, Z3 and Z4.

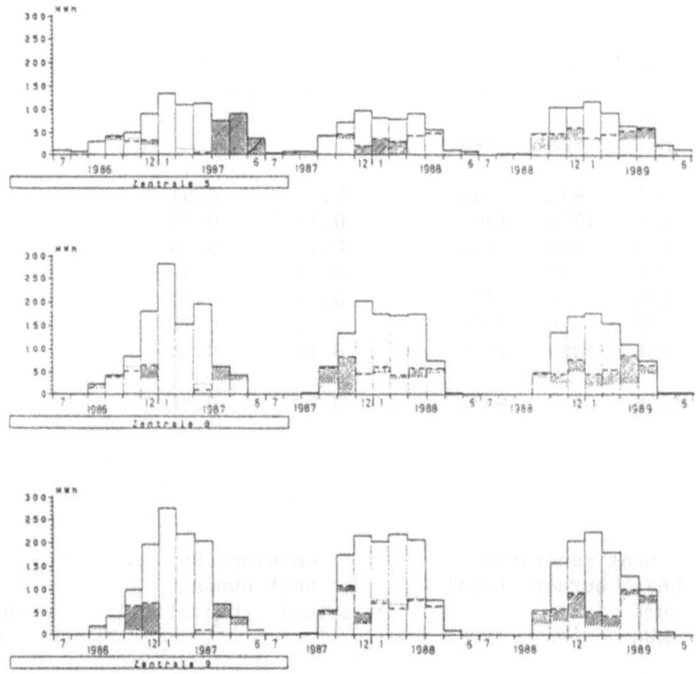

Fig. 5 b): Measured monthly energy consumption (solid lines) as well as actual (dotted lines) and theoretical (dashed lines) heat pump contributions from 1986 to 1989 for Z5, Z8 and Z9.

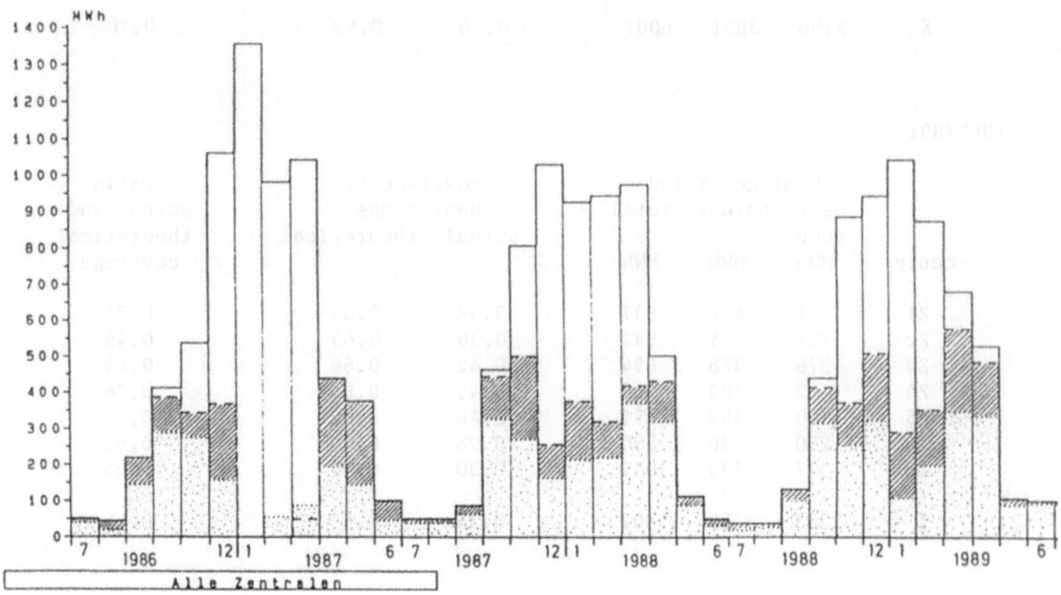

Fig. 5 c): Measured monthly energy consumption (solid lines) as well as actual (dotted lines) and theoretical (dashed lines) heat pump contributions from 1986 to 1989 for the sum of all these centers.

Especially positive results are observed for Z9 in the heating season 1987/88. The annual ratio of actual and theoretical contribution is 0.93 (Table 4).

1986/87:

center	heat generated: heat pump MWh	burner MWh	total MWh	coverage by heat pumps: actual	theoretical	ratio actual and theoretical coverage
Z1	341	622	963	0.35	0.41	0.86
Z2	135	1070	1205	0.11	0.39	0.29
Z3	175	458	633	0.28	0.38	0.73
Z4	248	548	796	0.31	0.39	0.79
Z5	175	612	787	0.22	0.45	0.49
Z8	238	828	1066	0.22	0.28	0.81
Z9	156	1029	1185	0.13	0.28	0.47
Σ	1468	5167	6635	0.22	0.36	0.61

1987/88:

center	heat generated: heat pump MWh	burner MWh	total MWh	coverage by heat pumps: actual	theoretical	ratio actual and theoretical coverage
Z1	395	480	875	0.45	0.58	0.78
Z2	329	655	984	0.33	0.57	0.58
Z3	286	381	667	0.43	0.58	0.74
Z4	218	522	740	0.30	0.57	0.52
Z5	200	354	554	0.36	0.56	0.65
Z8	256	752	1008	0.25	0.42	0.61
Z9	472	707	1179	0.40	0.43	0.93
Σ	2156	3851	6007	0.36	0.52	0.70

1988/89:

center	heat generated: heat pump MWh	burner MWh	total MWh	coverage by heat pumps: actual	theoretical	ratio actual and theoretical coverage
Z1	401	436	837	0.48	0.65	0.74
Z2	297	685	982	0.30	0.65	0.46
Z3	276	378	654	0.42	0.66	0.64
Z4	362	383	745	0.49	0.64	0.76
Z5	300	352	652	0.46	0.62	0.74
Z8	250	630	880	0.28	0.49	0.58
Z9	317	735	1052	0.30	0.47	0.65
Σ	2203	3599	5802	0.38	0.59	0.65

Table 4: Heat delivered by the heat pumps and burners in absolute and relative numbers for three heating seasons.

5.3 Measured Demand And Power Rating

In Table 5 the measured maximum demand for heating and hot water ($T_{outdoor}$= -15°C) is compared with the installed power of the burners. One recognizes that the actual demand is about 30% less than the installed power. In Z5 the mismatch is more than 100%. In Table 6 the equivalent demand at +3°C (switching temperature for the heat pumps) is compared to the rated power of the heat pumps. Here a similar situation is observed, except for Z8 and Z9, where the rated power was purposely chosen below the demand.

center	hot water demand kW	heating demand (TU=-15°C) kW	total demand (TU=-15°C) kW	burner rated power kW	ratio of rated power to total demand
Z1	20	390	410	570	1.39
Z2	35	440	475	650	1.37
Z3	20	300	320	425	1.33
Z4	20	330	350	496	1.42
Z5	30	310	340	730	2.15
Z8	—	640	640	815	1.27
Z9	—	660	660	896	1.36

Table 5: Heating demand for $T_{outdoor}$=-15°C and rated power of burners.

center	hot water demand kW	heating demand (TU=+3°C) kW	total demand (TU=+3°C) kW	burner rated power kW	ratio of rated power to total demand
Z1	20	185	205	290	1.41
Z2	35	200	235	325	1.39
Z3	20	145	165	196	1.19
Z4	20	160	180	229	1.27
Z5	30	135	165	400	2.42
Z8	—	275	275	173	0.63
Z9	—	275	275	220	0.80

Table 6: Heating demand for $T_{outdoor}$=+3°C and rated power of heat pumps.

5.4 Efficiencies And Coefficients Of Performance

An important quality for the characterization of a heat supply system is the ratio of heat output and fossile primary energy input; for burners this ratio is given by the efficiency η_K, for heat pumps by the coefficient of performance (cop) η_W. In this report we only present data averaged over one year.

In Table 7 efficiencies η_K for the years 1988 and 1989 are given. The spread in η_K from 0.79 to 0.88 results either from a non-optimal burner adjustment (Z1) or from summer operation with especially short on-periods (compare Z4 in 1988 and 1989).

year	Z1	Z2	Z3	Z4	Z5	Z8	Z9
1988	0.79	0.81	0.85	0.79	0.80	0.82	0.84
1989	0.79	0.82	0.81	0.88	0.80	0.80	0.87

Table 7: Annual efficiency η_K of burners for 1988 and 1989.

In Table 8 the cop η_W is presented for the years 1988 and 1989. Especially negative appears the value $\eta_W \approx 1.05$ for Z5. The reason very likely is a non-optimal gas/air mixture. A considerable improvement from η_W=1.10 to 1.40 could be observed in April 1988 for Z1, where a new mixer was installed (see Fig. 6). In Z2 a low annual cop is caused by problems in the control system;

in periods with regular operation $\eta_w \approx 1.40$ is determined. The best annual cops of 1.48 and 1.56 are observed in Z9 with an extraordinary large anergy collector. Values of 1.40 to 1.45 as evaluated for Z4 seem to be realistic for optimal heat pump operation in our heating systems with outgoing fluid temperature of 55°C.

year	Z1	Z2	Z3	Z4	Z5	Z8	Z9
1988	1.26	1.27	1.32	1.45	1.09	1.36	1.48
1989	1.39	1.33	1.32	1.40	1.05	1.38	1.56

Table 8: Annual cop η_w of heat pumps for 1988 and 1989.

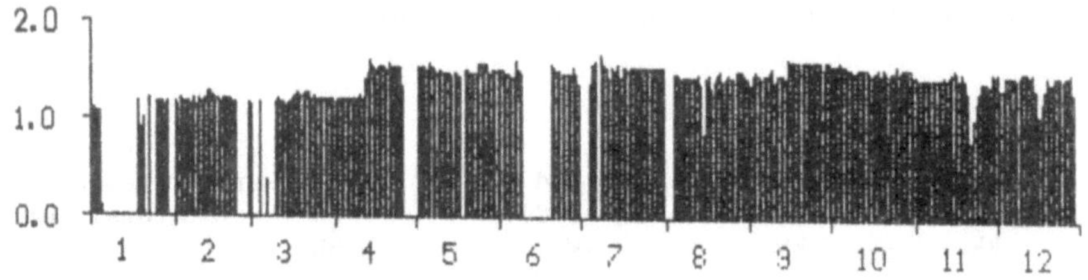

Fig. 6: Daily cop for heat pump in Z1 for 1988.

The variation of cop with temperature of incoming collector fluid is shown in Fig. 7. Data have been selected for stationary heat pump operation (on-phases longer than 60 minutes). The cloud of data points represents Z3. A general feature is the linear increase of the cop with rising temperature, indicated by straight lines. The improvement in Z1 operation is not effective, yet.

In contrast to the burner operation, where the seasonal η_K is larger in winter months, we observe no significant change of cop with the daily demand. From Fig. 7, however, one would expect an increase of cop in summer. This effect might be compensated by increased losses occurring during periods with short daily operating times.

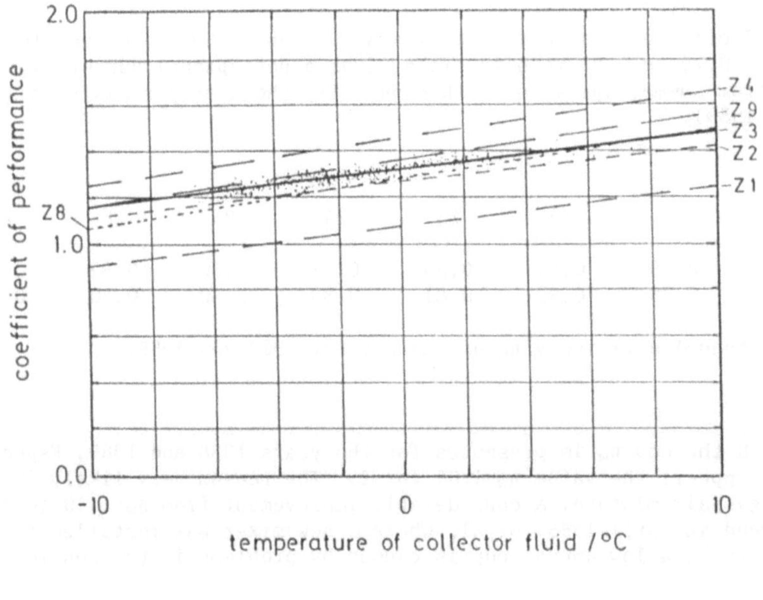

Fig. 7: Cop vs. temperature of incoming collector fluid between February and April 1988; the data points and the solid line are for Z3, the dashed lines for Z1, Z2, Z4, Z8 and Z9.

5.5 Savings On Primary Energy

The main question of the Volkach project is the evaluation of the realistic primary energy saving potential.

The heating demand QHT is provided by heat pump (fraction w) and burner (fraction 1-w). This causes primary energy inputs for bivalent operation

$$E_W = w \cdot QHT / \eta_W \quad \text{and}$$

$$E_K = (1-w) \cdot QHT / \eta_K .$$

This has to be compared with

$$E_K' = QHT / \eta_K'$$

for the hypothetical case of pure burner operation with annual efficiency η_K'. As mentioned above one expects $\eta_K'<\eta_K$ due to summer operation.

The absolute amount of primary energy saved is

$$S = E_K' - (E_W + E_K)$$

$$= [(1/\eta_K'-1/\eta_K) + (1/\eta_K-1/\eta_W) \cdot w] \cdot QHT .$$

The relative saving therefore is

$$s = S / E_K'$$

$$= [(1/\eta_K'-1/\eta_K) + (1/\eta_K-1/\eta_W) \cdot w] \cdot \eta_K' .$$

If one chooses $\eta_K'=0.79$ (which is the smallest measured annual efficiency for Z4 with extended summer operation in 1988), and takes the measured values for η_K, η_W, w and QHT for the three investigated heating seasons (see Tables 4, 7 and 8), one gets the actual absolute and relative primary energy saving S_{act} and s_{act} (Table 9).

If we put $\eta_K=\eta_K'$ and $\eta_W=1.45$ (realistic for optimal heat pump operation) and assume the maximum theoretical coverage by the heat pumps (see Table 4), one gets the maximum theoretical absolute and relative primary energy saving S_{theo} and s_{theo} (Table 9).

Our investigations showed that in 1986/87 about 900 MWh (11%), in 1987/88 1200 MWh (16%) and in 1988/89 1200 MWh (16%) of fossile primary energy was saved. this corresponds to 90 000, 120 000 and 120 000 l of oil.

For optimal heat pump operation without malfunctions and failures in 1986/87 1400 MWh (16%), in 1987/88 1800 MWh (24%) and in 1988/89 2000 MWh (27%) could have been saved. Thus we find, that for all three heating seasons only two thirds of the possible savings have been realized.

Unfavourable results, documented in a small ratio $x=S_{act}/S_{theo}$, are observed for

- Z1, where only 54% and 49% of the possible savings were realized in 1986/87 and 1987/88; this was due to the small cop,
- Z2, with x=36%, 54% and 41% due to problems in the control system,
- Z4, with x=51% in 1987/88 due to failure of the heat pump,
- Z5, with x=34%, 42% and 33% due to a small cop and to longer and frequent off-times, which were caused by a damaged gas leak detector of the safety system, and

- Z8, with x=40% in 1988/89 due to non-professional changes in the system with negative consequences for controlling.

Positive results are observed for

- Z3, with x=94% in 1986/87,
- Z8, with x=95% in 1986/87, and
- Z9, with x=88% in 1986/87, x=114% in 1987/88 and x=92% in 1988/89; in 1988/89 the very large cop compensates for irregular operation introduced by similar changes in the system as for Z8.

The reason for x>100% in Z9 is a consequence of the special choice of η_K in the calculation of S_{theo}.

The variation of the possible relative savings s_{theo} (13% to 20% in 1986/87, 19% to 26% in 1987/88 and 21% to 30% in 1988/89) is caused by the power rating of the heat pumps and the meteorological situation in the heating seasons. A larger average ambient temperature increases the possible savings from heat pump operation.

1986/87:

center	actual saving absolute S_{act}/MWh	actual saving relative s_{act}	theoretical saving absolute S_{theo}/MWh	theoretical saving relative s_{theo}	ratio x S_{act}/S_{theo}
Z1	123	0.10	227	0.19	0.54
Z2	97	0.06	271	0.18	0.36
Z3	131	0.16	139	0.17	0.94
Z4	142	0.14	179	0.18	0.79
Z5	70	0.07	204	0.20	0.34
Z8	163	0.12	172	0.13	0.95
Z9	169	0.11	191	0.13	0.88
Σ	894	0.11	1383	0.16	0.65

1987/88:

center	actual saving absolute S_{act}/MWh	actual saving relative s_{act}	theoretical saving absolute S_{theo}/MWh	theoretical saving relative s_{theo}	ratio x S_{act}/S_{theo}
Z1	144	0.13	292	0.26	0.49
Z2	176	0.14	323	0.26	0.54
Z3	180	0.21	223	0.26	0.81
Z4	124	0.13	243	0.26	0.51
Z5	75	0.11	179	0.25	0.42
Z8	169	0.13	244	0.19	0.69
Z9	332	0.22	292	0.20	1.14
Σ	1198	0.16	1796	0.24	0.67

1988/89:

center	actual saving absolute S_{act}/MWh	actual saving relative s_{act}	theoretical saving absolute S_{theo}/MWh	theoretical saving relative s_{theo}	ratio x S_{act}/S_{theo}
Z1	210	0.20	313	0.30	0.67
Z2	152	0.12	368	0.30	0.41
Z3	146	0.18	249	0.30	0.59
Z4	232	0.25	275	0.29	0.84
Z5	77	0.09	233	0.28	0.33
Z8	100	0.09	248	0.22	0.40
Z9	263	0.20	285	0.21	0.92
Σ	1180	0.16	1971	0.27	0.60

Table 9: Primary energy savings in three heating seasons.

6. OUTLOOK

In order to realize the largest possible saving of primary energy a number of measures has to be performed:

- debugging of the control system and optimization,
- improvement of the cop by optimal adjustment of the gas/air mixture, and
- lowering of the switch-off temperature for the heat pumps.

With respect to the last point, Fig. 8 depicts the temperature-duration curve, which indicates the number of days or hours with ambient temperature staying below a certain value. The hatched and cross-hatched areas represent the quantity "degree-days", which is proportional to the heating demand (excluding hot water demand). The cross-hatched area is indicative for the fraction, which cannot be provided by the heat pumps ($T_{outdoor}$ < +3°C, see Fig. 8 a). The fraction covered by the heat pumps thus ought to be 59%. If the switch-off temperature were shifted downward from 3°C to 0°C (Fig. 8 b) the heat pump fraction could be increased to about 90%.

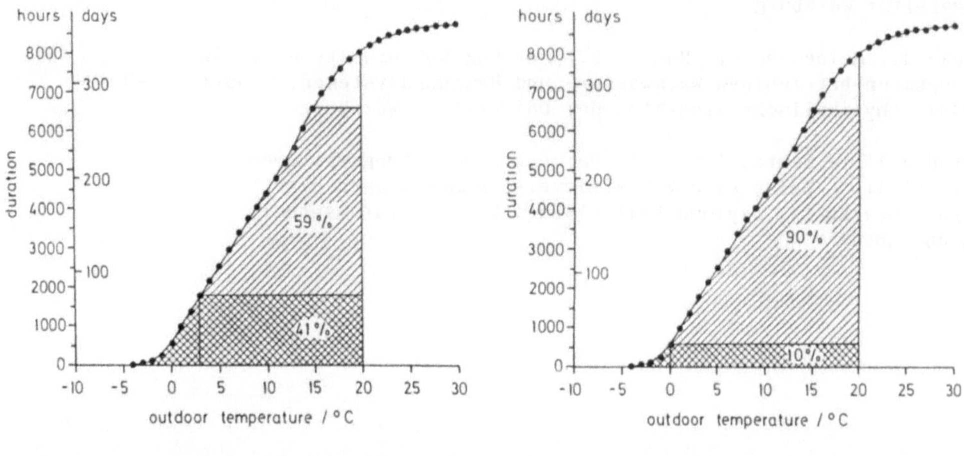

Fig. 8: Temperature-duration curve for $T_{outdoor}$ in 1988 a) for a switch-off temperature of +3°C, b) of 0°C; the hatched and cross-hatched areas indicate the total heating demand, the cross-hatched area is equivalent to the demand, which cannot be covered by the heat pumps.

In order to estimate an upper limit for the saving of primary energy, we assume a heat pump coverage of 100% (w=1.00). A relative amount of primary energy

$$s = 1 - \eta_K'/\eta_W$$

would be saved under these conditions. With $\eta_K'=0.79$ and $\eta_W=1.45$ we get s=0.46.

In absolute numbers this corresponds to savings of 3300 MWh or 330 000 l of oil for 1988/89.

Though this appears to be rather attractive, especially with respect to CO_2 output, it should be mentioned that in general gas burners operate with less emissions of CO, NO_x and carbohydrates than gas engine driven heat pumps if equal amounts of produced heat are considered. The CFC emissions from heat pumps are an additional problem. Catalytic systems may reduce pollution problems, but up to now no scrap devices are installed in Volkach.

We would like to point out that the thermal insulation of these barracks built in the middle of the eighties is provided by a double wall construction with a U-value of 0.5 $W/(m^2 \cdot K)$ and triple glazed windows. Thus the energy demand of 90 kWh per m^2 of living area and year is especially low compared to the german average of about 200 $kWh/(m^2 \cdot a)$.

ACKNOWLEDGEMENTS

The scientific research program is supported by the BMFT (German Federal Ministry for Research and Technology).

LITERATURE

Maderstein M.: Energiesammler als Wärmequellenanlage in Wärmepumpen-Heizungsanlegen, Report E12-0887-2 (1987), Diplomarbeit, Physikalisches Institut der Universität Würzburg

Burzynski B.: Untersuchungen von Kompaktabsorbern und Wärmepumpen, Report E21-0488-3 (1988), Diplomarbeit, Physikalisches Institut der Universität Würzburg

Fricke J., Heinemann U., Hümmer E.: Nutzung von Umweltenergie in Wärmepumpen-betriebenen Warmwasser- und Heizungssystemen, Report E21-0989-1 (1989), Physikalisches Institut der Universität Würzburg

Conrad R.: Die Rentabilität von Wärmepumpen-Heizungssystemen - dargestellt am Beispiel der Energieversorgungsanlagen einer Truppenunterkunft, Diplomarbeit (1989), Universität Mannheim (unpublished)

Chapter 16

PREDICTIVE CONTROL OF A HYBRID HEATING SYSTEM INCLUDING A HEAT PUMP

Christian Svensson (Norwegian Institute of Technology, Norway

Abstract. **This paper outlines the basic principle of a predictive control system, and shows how this can be used for controlling temperature set-back in buildings where a heat pump is an integrated part of the heating system.**

Because of the heat pump's distinctive character, it sets special requirements for the control system to save exergy. The predictive controller satisfies these requirements, and takes, unlike traditional bang-bang control, into account the heat pump's capability to conserve exergy.

Process knowledge in the form of a dynamic mathematical model is integrated in the controller, and an optimization algorithm finds the control trajectory that minimize the cost of exergy consumption. The Kuhn-Tucker conditions verify a strong active constraint which consists of a maximum utilization of the heat pump, and a minimum utilization of the auxiliary heater. This result can be used to simplify both the solution of the optimization problem, and the mathematical model.

Different control structures are compared regarding exergy consumption, and it is shown how these control structures influence the various exergy losses in the heating system.

A sensitivity analysis is performed, where different process parameters such as heat exchanger areas, thermal capacity of the building and part load efficiency of the compressor, are varied. It is shown how these perturbations influence the optimal control trajectory.

Introduction

Temperature set-back in buildings with a heat pump as an integrated part of the heating system, has been considered by several authors [2,3,6,12]. They observed that traditional control based on a fixed reduction of the thermostat set point, was not sufficient to achieve exergy saving. Both Harsem (1986) and Benton (1983) discovered in some cases a negative exergy saving. Benton explained that this was due to the use of a less efficient auxiliary heater for temperature recovery, which offsets savings accrued through temperature set-back. He proposed an improved control strategy that reduces or eliminates the use of the auxiliary heater, by letting the set point raise slowly for temperature recovery.

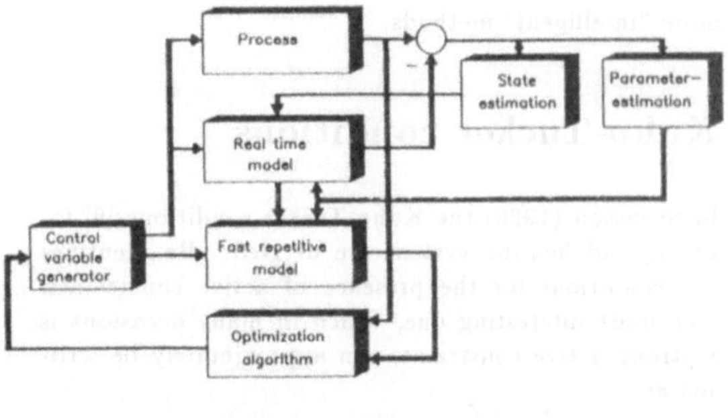

Figure 1: Block diagram of a predictive control system.

A predictive controller proves, in contradiction to fixed set point reduction and traditional bang-bang control, to take into account the heat pump's capability to conserve exergy. It will under any operating condition seek the control and temperature trajectory that minimize the cost of exergy consumption during the temperature recovery and set-back period.

Predictive control

Figure 1 shows the principle of a control structure that is called predictive control [1,5,10]. The system consists of a state and parameter estimation routine that works in real time. The parameters of the process model are continuously updated. In this way the model represents the process with sufficient accuracy. A model with exactly the same structure and with updated parameters, but with a much smaller time scale, is made to produce responses to fast repetitions of different control actions. The optimization algorithm seeks the control trajectory that mimimizes the optimum criterion.

Heat Pumps – Proceedings of the 4th International Conference, pp.177-182

It is difficult to find a solution of this dynamic optimization problem because of the combination of a non-linear process model and a two point boundary value problem. Because of this it is often convenient to convert the dynamic optimization problem to a non-linear optimization problem. This is done by anticipating certain restrictions on the control trajectory. Here the method of discretization of the control function is used [4].

$$\mathbf{u}(t) : \mathbf{u}(0), \mathbf{u}(1), \ldots, \mathbf{u}(n-1) \tag{1}$$

The non-linear optimization problem can be solved in different ways. In this paper we use an adapted coordinate method,which makes it capable of handling constraints. In each step the method finds the coordinate direction with the largest negative gradient component of the objective function. See Svensson (1990) for details.

The method can handle suboptimums, it is robust and easy to implement, but it is more time consuming than more "intelligent" methods.

Kuhn-Tucker conditions

In Svensson (1990) the Kuhn-Tucker conditions [9] for the hybrid heating system are derived. He identifies the conditions for the presence of active constraints. The most interesting one, which in many occasions is a strong active constraint, can approximately be written as:

$$\frac{\zeta_{aux}}{\zeta_{hp}} \cdot \frac{c_{hp}}{c_{aux}} \leq 1.0 \tag{2}$$

Here ζ_{aux} and ζ_{hp} are the mean exergy efficiency of the auxiliary heater and the heat pump. c_{hp}/c_{aux} is the ratio of the unit cost of input exergy.

When the constraint (2) is active, it means that the optimal control strategy during the set-back period is to maximize the utilization of the heat pump, and to minimize the utilization of the auxiliary heater.

An active constraint can be used to generate simplifications in the predictive controller. There is no need for the optimization algorithm to find the optimal exergy allocation between the auxiliary heater and the heat pump. The problem reduces to the solution of a two point boundary value problem.

In the following we do not take advantage of this observation, but use the predictive controller as originally described.

Example

The simulated heating system is shown in Figure 2. It consists of a heat pump, an oil-fired boiler, a heat distribution system and a building. The heating system is characterized with its significant differences in the eigenvalues, and this is taken advantage of in the model. See Stoecker (1975).The heat pump, boiler and heat distribution system are modelled as a set of steady-state equations. The building is modelled with differential equations to preserve its dynamic characteristic. The energy balance of the building is:

$$\frac{d}{dt}\iiint_V \rho E dV + \iint_A \rho v E dA = \dot{Q} \tag{3}$$

Se Thorbergsen (1985), Svensson (1990) and Appendix for more details about the model.

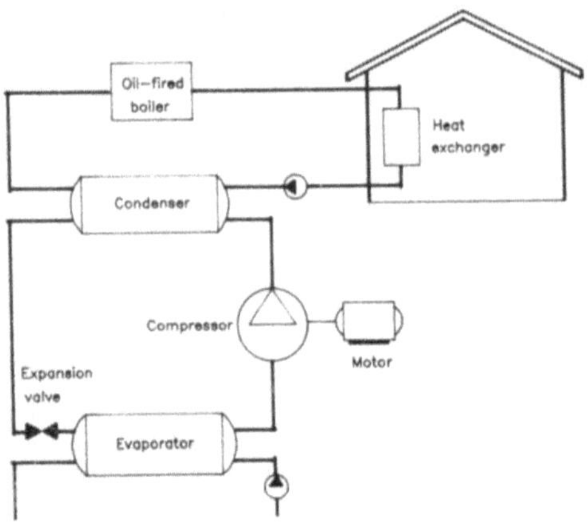

Figure 2: Flow chart of a hybrid heating system.

For detailed information on the state and parameter estimator, we refer to an enormous litterature. See Hays (1979), Ljung (1985) and Sorenson (1985).

The optimization problem consists of finding the control trajectory that minimizes the cost of exergy consumption of the heating system in the actual time span (16 hours).

Minimize

$$J = \int_{t_0}^{t_1} (c_{el} P_{motor} + c_{el} P_{pumps} + c_{oil} P_{boiler}) dt \tag{4}$$

subject to

$$\dot{\mathbf{x}} = f(\mathbf{x}, \mathbf{u}, t) \tag{5}$$

$$x_{build}(t_0) = 20.0°C \tag{6}$$

$$x_{build}(t_1) = 20.0°C \tag{7}$$

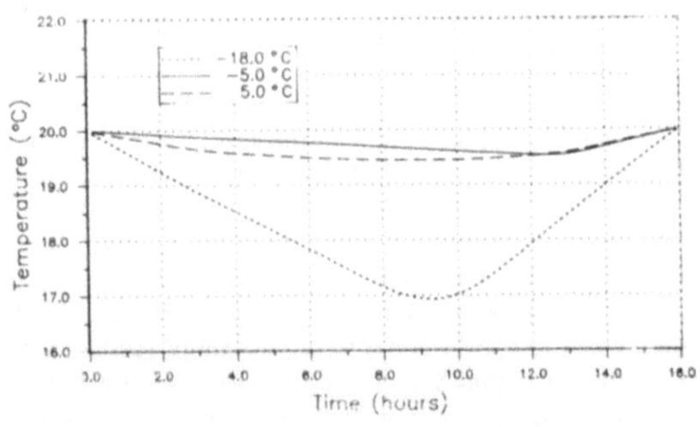

Figure 3: Optimal temperature trajectory in the building with three different operation conditions, $T_{amb} = -18.0°C$, $T_{amb} = -5.0°C$ and $T_{amb} = 5.0°C$.

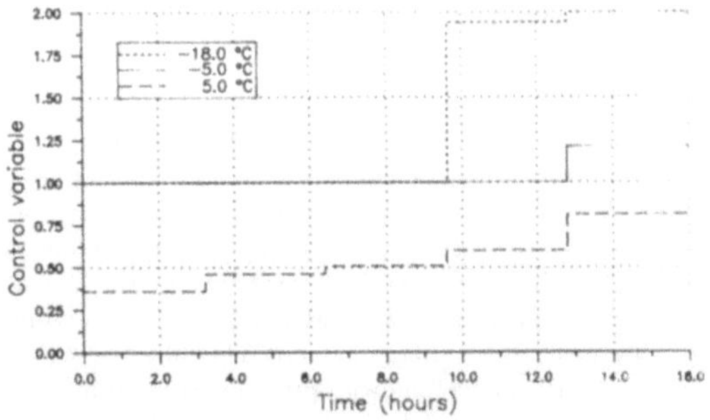

Figure 4: **Optimal** control trajectory **with three** different operation **conditions. 0.0-1.0: Relative** compressor capacity. 1.0-2.0: Relative boiler capacity.

Results

Figure 3 and Figure 4 show the optimal control trajectory and temperature trajectory with three different operation conditions. In Table 1 a comparison is made between the predictive controller and the traditional bang-bang controller. A control system that does not utilize the possibility for temperature set-back is used as a reference. Bang-bang control means that the heating system "bangs" off at the end of occupancy,t_0, and "bangs" on with the control variable at its maximum just in time to fulfil the temperature request at the beginning of occupancy,t_1. The ration of cost of exergy, c_{el}/c_{oil}, is 2.0.

At heavy load, $T_{amb} = -18°C$, the predictive controller uses the reduced heat demand to minimize the use of the boiler due to its low exergy efficiency. This is a comfirmation of the inequality (2). It is a well known fact that the boiler uses exergy to cover both the exergy and the anergy part of its heat delivery. The predictive controller reduces the exergy consumption with 10.1% (within the 16 hours time span) compared to no set-back.

Operating condition	Control strategy	Control trajectory[a]	Exergy transfer[b]	Exergy losses	Exergy input el	Exergy input oil
High load (−18.0°C)	*Predictive*	[1.00 1.00 1.01 1.94 2.00]	210.0[c]	914.6	281.6	781.9
	No set-back	[1.45 1.45 1.45 1.45 1.45]	228.6	1014.1	290.8	892.5
	Bang-bang	[0.00 0.00 2.00 2.00 2.00]	198.6	1281.2	195.0	1250.0
Average load (−5.0°C)	*Predictive*	[1.00 1.00 0.99 1.00 1.21]	97.0	279.5	266.8	86.5
	No set-back	[1.05 1.05 1.05 1.05 1.05]	98.9	295.5	268.3	103.1
	Bang-bang	[0.00 0.00 0.00 1.96 1.96]	78.4	825.1	123.9	771.2
Low load (5.0°C)	*Predictive*	[0.36 0.46 0.51 0.60 0.81]	33.9	108.1	142.0	-
	No set-back	[0.56 0.56 0.56 0.56 0.56]	35.6	109.0	144.6	-
	Bang-bang[d]	[0.00 0.00 0.98 0.98 0.98]	30.4	123.1	153.5	-

[a]0.0 - 1.0: Relative compressor capacity, 1.0 - 2.0: Relative boiler capacity.

[b]Exergy transfer to the building.

[c]All figures in kWh.

[d]Only the heat pump is bang-bang controlled.

Table 1: Comparison of different control **strategies** with three different operating **conditions.**

Note the increased exergy consumption when the traditional bang-bang controller is used (22.1%). In this case no advantage is taken of the anergy and exergy content of the heat source when the heat pump is turned off.

At average load, $T_{amb} = -5°C$, the predictive controller still utilizes the heat pump's good exergy efficiency. Just a minor part of the exergy saving-potensial of the building is utilized. The main target is still to avoid the use of the boiler. The exergy consumption is reduced by 4.9%. As seen from Figure 3 a suboptimal solution is generated because of the constrained control function. By decreasing the discretization time the solution will converge to the optimal dynamic solution.

At low load, $T_{amb} = 5°C$, the optimal control strategy is to control the heat pump at part load. The exergy consumption is reduced with 1.8% compared with no set-back. Figure 5 shows the distribution of the various exergy losses in the heating system with optimal control. Figure 6 shows the same losses when the heat pump is bang-bang controlled. The losses associated with a fast temperature recovery (bang-bang control) is bigger then the part-load and pumping losses with predictive control. This is mostly due to a higher temperature lift and a poorer utilizing of the heat exchanger areas during the recovery period.

The total saving of exergy during a heating season is 8.5%. The main contribution comes from the reduced input exergy to the oil-fired boiler .

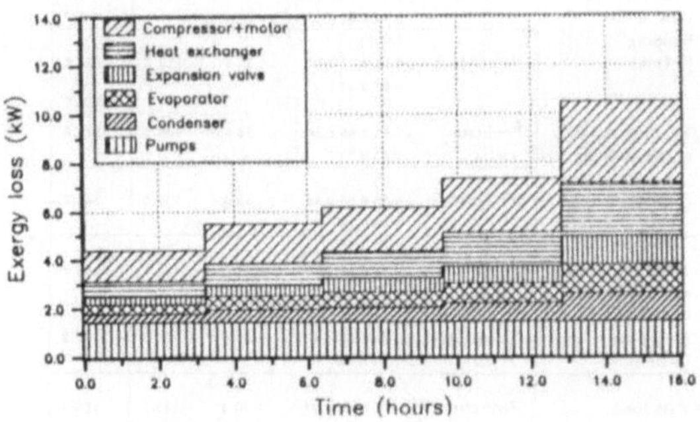

Figure 5: The distribution of the various exergy losses in the heating system with predictive control. $T_{amb} = 5°C$.

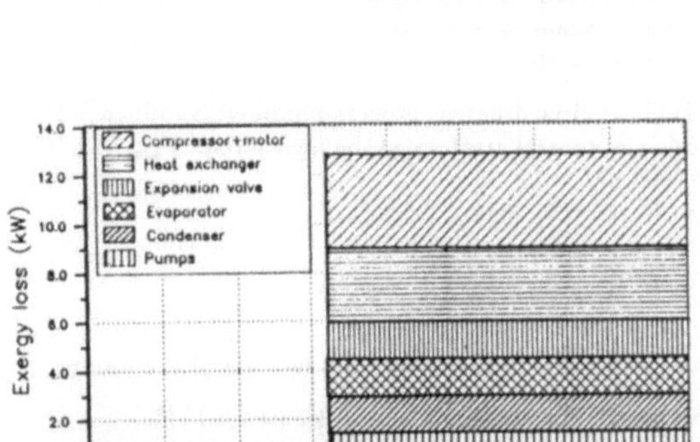

Figure 6: The distribution of the various exergy losses in the heating system with bang-bang control. $T_{amb} = 5°C$.

Sensitivity analysis

The purpose of the sensitivity analysis is to illustrate how different process parameters influence the optimal control trajectory of the heat pump. The analysis is performed at low load, $T_{amb} = 5°C$, and the results are given in Table 2.

A reduction of the heat exchanger area in the building (25%) results in an increased exergy consumption (7%), but an insignificant change of the optimal control strategy. The increased exergy losses are mostly design dependent, and the controller can do nothing to reduce them.

An increased thermal capacity of the building (28%) results in a reduced exergy saving. The optimal control strategy approaches the *no set-back* strategy. The increased time constant of the building results in a higher temperature during the set-back period. The losses during the recovery period become relatively more significant and is reduced.

An increase in the mass flow rates (20%) on both the low and high temperature side result in a turn-off early in the period. The pumping losses become more dominant

Process parameters[a]	Control strategy	Control trajectory[b]	Exergy transfer	Exergy losses	Exergy input
Heat exchanger area in the building (−25%)	*Predictive*	[0.37 0.47 0.52 0.62 0.82]	33.9[c]	118.0	151.9
	No set-back	[0.57 0.57 0.57 0.57 0.57]	35.6	119.2	154.8
Thermal capacity of the building (+28%)	*Predictive*	[0.41 0.46 0.56 0.61 0.73]	34.6	108.4	143.0
	No set-back	[0.55 0.56 0.56 0.55 0.56]	35.6	109.0	144.6
Mass flow rates of water on the low and high temperature side (+20%)	*Predictive*	[0.00 0.48 0.58 0.63 0.99]	31.7	121.1	152.8
	No set-back	[0.54 0.54 0.54 0.54 0.54]	35.6	121.6	157.2
Part load efficiency of the compressor (Valve unloading)	*Predictive*	[0.00 0.00 0.97 0.97 0.97]	30.4	123.7	154.1
	No set-back	[0.54 0.54 0.54 0.54 0.54]	35.6	129.7	165.3

[a]The perturbation is given in parenthesis.

[b]0.0 - 1.0: Relative compressor capacity.

[c]All figures in kWh.

Table 2: How different process parameters influence the optimal control strategy of the heat pump, $T_{amb} = 5°C$.

and move the optimum towards less use of the pumps.

So far we have assumed an ideal part load efficiency for the compressor. This is replaced with a typical part load efficiency for a piston compressor with valve unloading. See Nowotny (1980). The optimal control strategy is now bang-bang control of the heat pump. The predictive controller avoids part load operation because of increased part load losses.

Conclusions

The predictive controller proves to have sufficient degree of freedom to achieve considerable exergy saving in temperature set-back control of hybrid heating systems. Most of the saving is a result of obtaining an optimal balance between the heat pump and the auxiliary heater. This balance is in most cases characterized by a maximum utilization of the heat pump and a minimum utilization of the auxiliary heater. This observation is identical with the wellknown optimal steady-state control strategy of hybrid heating systems including heat pumps.

References

[1] Balchen, J.G. and Mummé, K.I. (1988). *Process Control. Structures and Applications.* Van Nostrand Reinhold Co.,New York.

[2] Benton, R. (1983). Heat Pump Setback: Computer Prediction and Field Test Verification of Energy Savings with Improved Control. *ASHRAE Transactions 89:Ib 1983.*

[3] Bullock, C.E. (1978). Thermostat Setback and Residental Heat Pumps. *ASHRAE Journal, sept. 1978.*

[4] Edgar, T.F. and Himmelblau, D.M. (1988). *Optimization of Chemical Processes.* McGraw-Hill Book Co.,New York.

[5] Garcia, D.M. et all (1989). Model Predictive Control: Theory and Practice - a Survey. *Automatica Vol.25 No.3 1989.*

[6] Harsem, T.T. (1986). Temperature set-back and its influence on the operation of heat pumps. *SINTEF-rapport nr. STF A86047. Trondheim (Norway).*

[7] Hayes, D.K. (1979). Real Time Identification of Parameters in Building Models. *International Congress Joint Meeting with the American Section of the I.S.E.S., Atlanta, Georgia. May 1979.*

[8] Ljung, L. and Söderström, T. (1985). *Theory and Practice of Recursive Identification.* The MIT Press, Cambridge, Massachusetts.

[9] Luenberger, D. G. (1984). *Linear and Nonlinear Programming.* Addison-Wesley Pub. Co., Reading, Massachusetts.

[10] Morari, M. (1988). Advances in Process Control Theory. *Chem. Eng. Prog. 10 (1988).*

[11] Nowotny, S. (1980). *Saving of Energy in Refrigeration.* International Institute of Refrigeration,Paris.

[12] Schade, G.R. (1978). Saving of Energy by Night Setback of a Residental Heat Pump System. *ASHRAE Transactions 84:I (1978).*

[13] Sorenson, H.W. (1985). *Kalman Filtering:Theory and Application.* IEEE PRESS,New York.

[14] Stoecker, W.F. (1975). Procedures for Simulating the Performance of Components and Systems for Energy Calculations. *ASHRAE Task Group on Energy Requirements for Heating and Cooling of Buildings.*

[15] Svensson, M.C. (1990). *Optimization and Identification of Heat Pumps.* Ph.D. Thesis. The Norwegian Institute of Technology. To be published.

[16] Thorbergsen, E. (1985). PROSIM, a Program System Simulating Heat Pump Plants. *IIR Commission E2 Trondheim (Norway) 1985.*

[17] Villadsen, V. (1985). Reciprocating Compressors for Refrigeration and Heat Pump Application. *Int. J. Refrig. Vol 8 September 1985.*

Notation

A	area of the control volume
A_{build}	outside area of the building
A_{ex}	heat exchanger area
$C_{p,build}$	mean specific heat capacity of the building
c_{aux}	price of auxiliary input exergy
c_{el}	price of electricity
c_{hp}	price of heat pump input exergy
c_{oil}	price of oil
E	specific energy
f	process model
J	objective function
k_{build}	overall coefficient of heat transfer, building
k_{cond}	overall coefficient of heat transfer, condenser
k_{evap}	overall coefficient of heat transfer, evaporator
k_{ex}	overall coefficient of heat transfer, heat exchanger
m_{build}	total mass of the building
$\dot{m}$	mass flow rate of water
$\dot{m}_{cond}$	mass flow rate of water in the condenser
$\dot{m}_{evap}$	mass flow rate of water in the evaporator
P_{boiler}	combustion heat input to the oil-fired boiler
P_{motor}	electrical power input to the motor
P_{pumps}	electrical power input to the pumps
P_{pump}	electrical power input to one pump
Q	heat transfer by conduction, building
$\dot{q}$	heat flow density in the evaporator
T_{amb}	ambient (outdoor) temperature
T_{build}	mean temperature of the building
ΔT_{lmtd}	logarithmic mean temp. difference,heat exchanger
t	time
t_0	end of occupancy
t_1	beginning of occupancy
$\mathbf{u}$	control variable (vector)
V	control volume
v	velocity vector
$\mathbf{x}$	state vector
x_{build}	mean temperature of the building
u	water flow velocity
α_{R22}	surface coeff. of heat transfer, refrigerant side
α_{water}	surface coeff. of heat transfer, water side
η_{boiler}	oil-fired boiler efficiency
η_{pump}	pump efficiency
ρ	density
ζ_{aux}	mean exergy efficiency of the auxiliary heater
ζ_{hp}	mean exergy efficiency of the heat pump

Appendix. Technical description of the hybrid heating system.

The heat source is sea water with constant temperature, 5.0^0C. The design heat load is 116 kW ($T_{amb} = -20.0^oC$), while the heat pump capacity is 65 kW.

At design heat load the supply/return temperature is $50.1^oC/39.1^oC$. The refrigerant is R22. PROSIM [16] is used to simulate the steady-state sub-system.

Evaporator.
The evaporator is a shell and tube evaporator with refrigerant-side-area 18.0 m^2. The heat transfer is characterized by the following equations:

$$\alpha_{R22} = 14.1\dot{q}^{0.565} \ , W/m^2K \tag{8}$$

$$\alpha_{water} = 3153.27u^{0.87} \ , W/m^2K \tag{9}$$

$$u = 0.6\dot{m}_{evap} \ , m/s \tag{10}$$

$$k_{evap} = \frac{1.0}{\frac{1.0}{\alpha_{R22}} + \frac{4.0}{\alpha_{water}}} \ , W/m^2K \tag{11}$$

Compressor.
The compressor is a reciprocating compressor with swept volume 0.021 m^3/s. Its volumetric and isentropic efficiency is a function of the pressure ratio and is given in Villadsen (1985). The part load efficiency for valve unloading is given in Nowotny (1980).

Motor.
The motor capacity is 20.0 kW. The motor efficiency is given in Nowotny (1980).

Condenser.
The condenser is a shell and tube condenser with refrigerant-side-area 14.0 m^2. The heat transfer is characterized by the following equations:

$$\alpha_{R22} = 1100.0 \ , W/m^2K \tag{12}$$

$$\alpha_{water} = 5662.0u^{0.87} \ , W/m^2K \tag{13}$$

$$u = 0.6\dot{m}_{cond} \ , m/s \tag{14}$$

$$k_{cond} = \frac{1.0}{\frac{1.0}{\alpha_{R22}} + \frac{1.6}{\alpha_{water}}} \ , W/m^2K \tag{15}$$

Oil-fired boiler.
The boiler capacity is 100.0 kW. The boiler efficiency is assumed constant:

$$\eta_{boiler} = 0.8 \tag{16}$$

Heat exchanger.
The heat exchanger area $A_{ex} = 160.0 \ m^2$. The overall coefficient of heat transfer is assumed constant:

$$k_{ex} = 30.0 \ W/m^2K \tag{17}$$

Pumps.
The pumps are identical, and it is assumed no heat losses. The following relations are used:

$$\eta_{pump} = 0.9 \tag{18}$$

$$P_{pump} = 48.0\dot{m}^3 \ , W \tag{19}$$

Mass flow rates of water on the low and high temperature side is:

$$\dot{m}_{evap} = 2.5 \ kg/s \tag{20}$$

$$\dot{m}_{cond} = 2.5 \ kg/s \tag{21}$$

Heat distribution system.
It is assumed no heat loss from the heat distribution system.

Building.
The following model and parameters are used to simulate the building:

$$\frac{d}{dt}(T_{build}) = \frac{1}{m_{build}C_{p,build}}[k_{ex}A_{ex}\Delta T_{lmtd} - k_{build}A_{build}(T_{build} - T_{amb})] \tag{22}$$

$$k_{build}A_{build} = 2.90 \ kW/K \tag{23}$$

$$m_{build}C_{p,build} = 39.15 \cdot 10^4 \ kJ/K \tag{24}$$

Chapter 17

FEEDING LARGE HEAT PUMPS FROM SEWAGE WATER TREATMENT PLANTS

B Stjernström (Flygt AB, Sweden)

SUMMARY

A number of large heat pump stations for district heating have recently been built in Sweden. Treated sewage water has been used as major low temperature energy source. Lake and sea water have also been used sometimes as additional energy to a sewage source. In the largest of them, the Hammarby 175 MW plant, water is transported from the sewage plant by gravity in a 2.7 km long rock tunnel and pumped up to and through the heat pump evaporators in two steps by submersible pumps. When returning to the recipient the head is used to recover energy with a 315 kW submersible hydroturbine generator. Multiple submersible pump concepts have been selected in order to cut total station costs and secure high availability. Basic pump station lay-outs are presented.

SWEDISH HEAT PUMP STATIONS

Since 1982 about 30 large heat pump plants for district heating purposes have been built in Sweden. The plants have between 7 to 120 MW thermal output power, the large ones divided into 20 to 30 MW units. The evaporators are of panel or tube type. Output temperatures from the heat pumps are between 50 to 80°C. Electric and/or oil fired boilers add the gap to sufficient distribution temperature and take care of peak loads. Large accumulators have been added to several stations, to even out peaks and take advantage of low cost electrical energy at night. For nearly all of them submersible pumps have been selected to pump the low temperature water from the source to the evaporators in the plants.

Energy Sources

Of course the higher the temperature of the source the less water has to be transported and the smaller the evaporators will be. Therefore treated sewage water and industrial waste water has, if available, been used as source. If not enough, lake or sea water has been added. Even lake or sea water alone has been used, but because of the low temperature two to three times more water has to pass through the evaporators.

TREATED SEWAGE WATER AS SOURCE

Sewage stations are often far away from heat pump stations. Therefore the treated sewage is usually pumped in steps. First from the

Heat Pumps – Proceedings of the 4th International Conference, pp.183-190

treatment plant to the location of the heat pump station with submersible centrifugal pumps. Then lifted with submersible propeller pumps over or through the condensers depending on type.
Treated sewage water normally has a temperature of 5 to 20°C. In this range 20 to 40 l/s is needed for each MW thermal output. As large heat pump units are between 20 to 30 MW output 400 to 1200 l/s is needed for each unit.

The Hammarby Heat Pump Station
In Hammarby the treated sewage water is taken through a 2.7 km long rock tunnel by gravity and then lifted 14 meters up to a distribution flume with speed regulated submersible centrifugal pumps. From the flume submersible propeller pump take the sewage water through the tube condensers.

Pump Selection 4 units each between 25 to 32 MW are switched in according to the demand, each using 700 l/s. They are fed from the flume individually with 4 submersible 110 kW propeller pumps working against 10 m head. From the gravity tunnel up to the distribution flume 3 submersible 220 kW centrifugal pumps deliver up to 3 x 960 l/s against a maximum head of 17 m. Two of the motors are speed controlled, the third runs at fixed speed. The Hammarby heat pump station is sized and pre-constructed for 3 more 32 MW units. Holes for 3 additional columns for submersible propeller pumps are already made as well as seats for additional lower submersible centrifugal pumps. One more unit has already been decided.

Choice of Pump Station The choice of multi-unit submersible pumps has been made because of high availability, safety and low total cost of the pump station where no dry pits or long shaft arrangements have to be made. Other advantages are easy access to the units for inspection, service or replacement. They are standard serial produced units with high spare part availability.

Energy Recovery Turbine The head drop from the flume to the outlet rock tunnel is about 10 m. Instead of just letting the sewage water fall down, and the energy be lost, a submersible hydroturbine generator has been installed in the vertical rock shaft. The turbine has automatically adjustable runner blades and operates today from 1400 l/s to 2800 l/s. The submersible hydroturbine generator is built for up to 4200 l/s available when the heat pump station has been extended to full capacity. At that time the submersible hydroturbine generator will generate 315 kW electrical power, returning around 30% of the energy demand for the submersible sewage water pumps.

Installation principle The turbine generator unit itself is standing by its own weight, and additional water pressure when running, on a bottom flange of a 10m steel pipe. The pipe is 1.4m in diameter and acts as a guide when lowering or lifting the submersible turbine generator. Power cables, control cable and flexible hoses for the hydraulically operated runner blades are guided on a rail inside the column and supported with a wire. A 9° straight conical draft tube recovers the dynamic head after the runner.

Cylinder Gate Control Starting and stopping the turbine is made by an oil hydraulically operated cylinder gate over the column inlet. The advantage with a cylinder gate is besides of a very smooth inflow, that it is hydraulically balanced. Therefore there is no need for accumulated back-up if the power supply fails. The gate closes automatically by gravity. This makes the oil hydraulic system very simple.

The stroke of the hydraulic piston is also shorter than for a sluice gate because the peripherical inlet area grows faster and the cylinder allows lower water level over the intake compared to a free surface without sucking air into the turbine.

Level Control Flume

Because of the limited space available and that the outlet from the distribution flume has a skiboard working as a flowmeter for the plant control, special arrangements had to be made to feed and control the submersible hydroturbine generator.

A small hydroflume was built in front of the cylinder gate after and just under the measuring skiboard. A second skiboard controls the maximum level and drops the overflow in the vertical rockshaft when the turbine generator is stopped. The automatically adjustable runner blades are controlled from the small flume surface level fluctuations maintaining almost constant level at all inflow conditions.

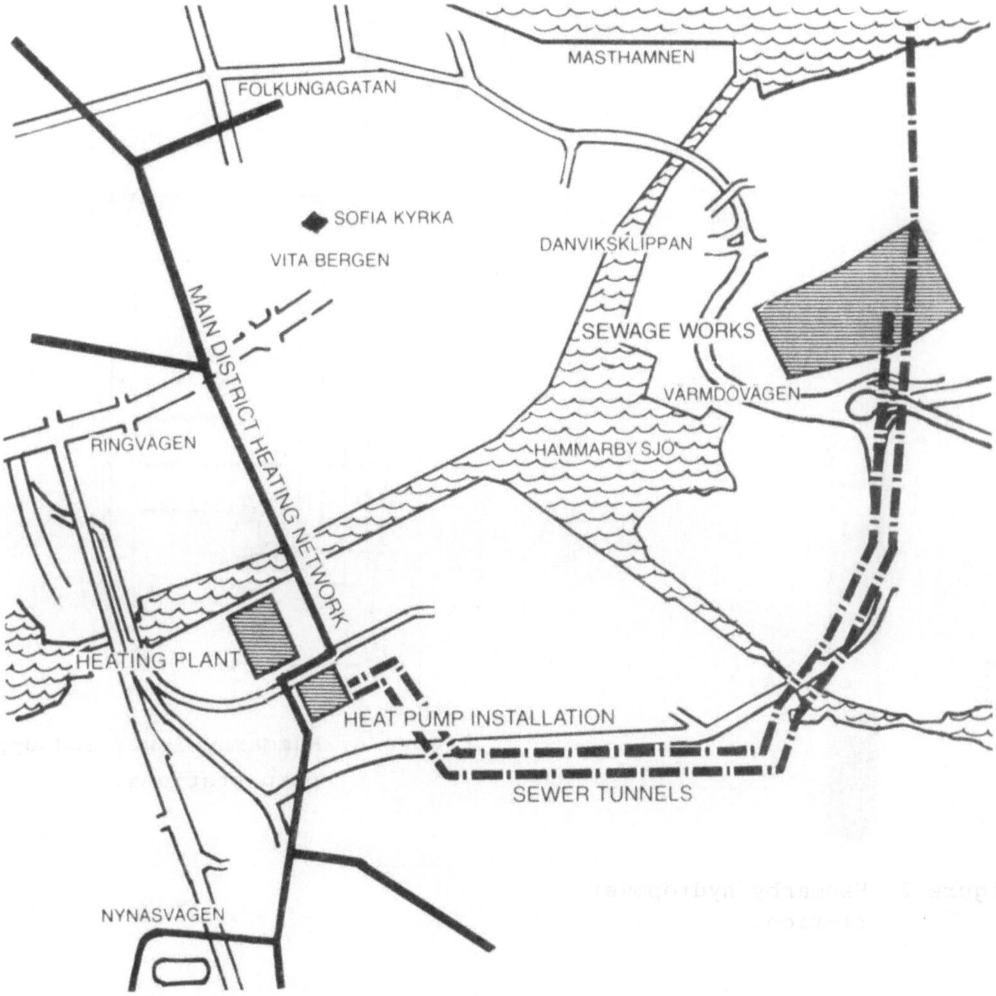

Figure 1. The Hammarby Plant - sewage source and district heating network.

LAKE AND SEA WATER AS SOURCE

Surface temperature of lake and seawater can reach up to 20°C during summertime. In winter, the surface of most Swedish lakes and costal waters is freezing up, but the bottom temperature in the sea and in deep lakes levels out at +4°C. Therefore a short summer intake is placed near the shore and a deep winter intake is located far out.

Lake or river water can be as cold as +1 to +2°C. This means that between 80 to 100 l/s is needed for each MW thermal power output. The large 20 to 30 MW units consume between 1600 h 3000 l/s each of lake river or seawater.

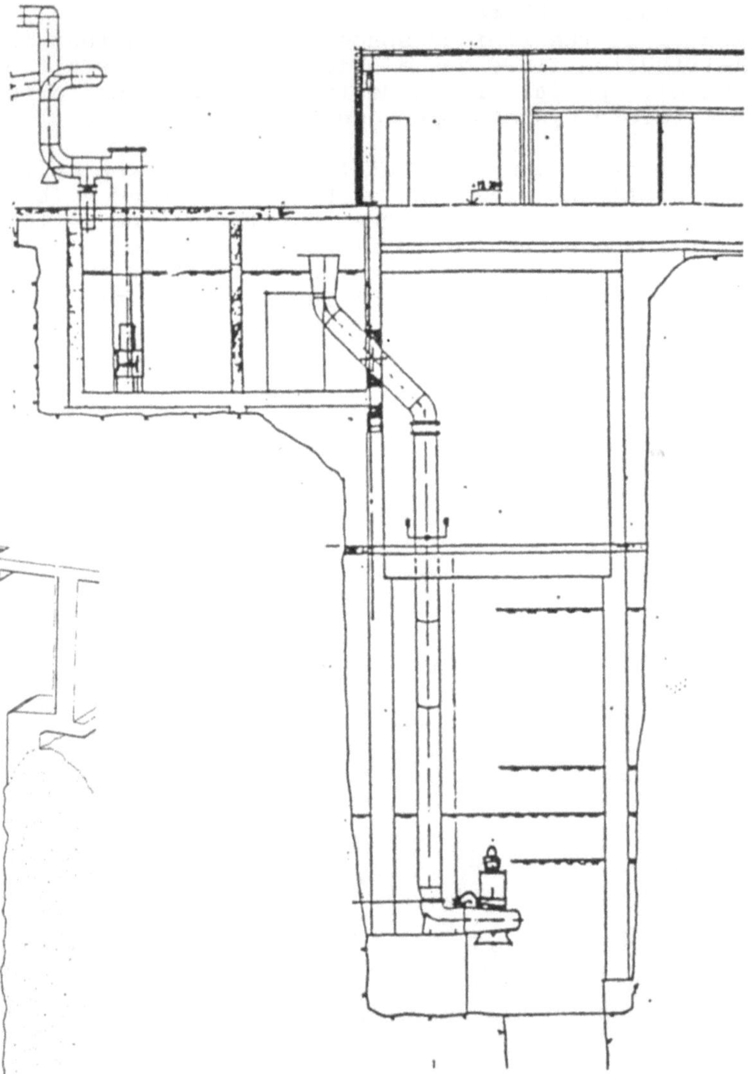

Figure 3. Hammarby lower and upper pump stations.

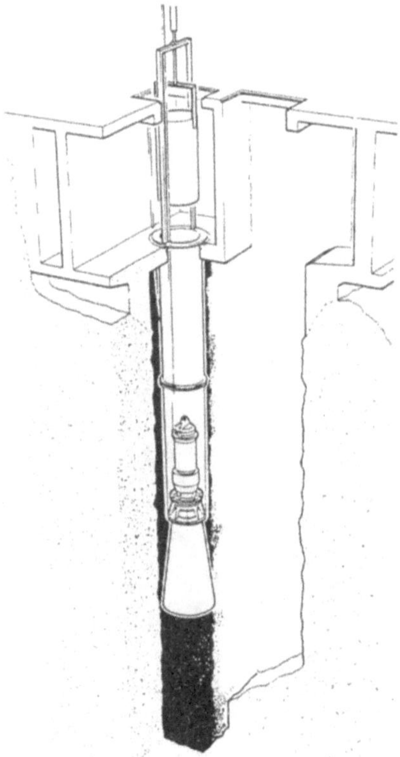

Figure 2. Hammarby hydropower station.

Basic lay-outs
Some pumping stations have been located directly at the shore and a second pump station at the heat pump station is taking the water to the evaporators in the same manner as with the sewage systems. In many cases the water is transported by gravity in a rock tunnel to submersible pumps under the heat pump station. After the evaporators the water is returned through a pipe inside the inflow rock tunnel, back to the lake or sea.

The winter intake pipes can be several kilometres long. Woodstave pipes have successfully been used. The long piping system have been carefully surge calculated.

Mixed sources

When there is not enough sewage water it can be mixed with lake or seawater. In Visby the sewage treatment plant supplies 20% of the flow and 80% is added from the sea below the heat pump station. A submersible energy recovery turbine generator is considered after the evaporators.

1. *Summer intake*
2. *Winter intake*
3. *Lake water pump*
4. *Low temperature evaporator*
5. *Turbo compressor*
6. *Electric motor*
7. *Gear*
8. *Condenser*
9. *Condensate cooler*
10. *Shut-off valve*
11. *District heating pump*
12. *Shut-off valve*
13. *High pressure control valve*
14. *Flash box*
15. *Low pressure control valve*

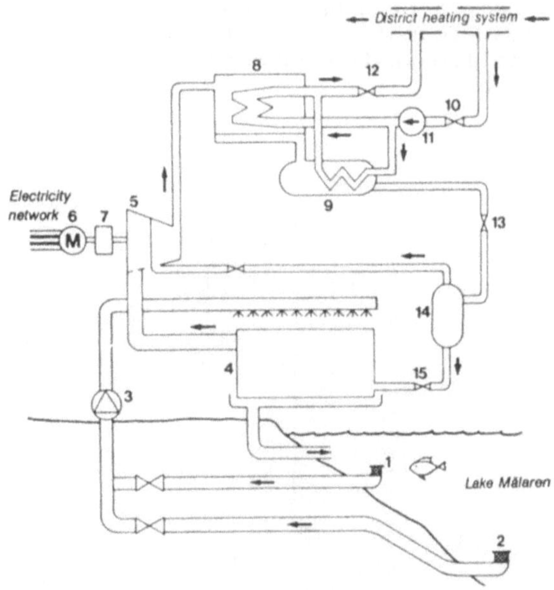

Figure 4. Lake or seawater as source - summer and winter intakes.

THE SUBMERSIBLE MACHINERY

A submersible pump or hydroturbine generator is designed in the same basic way. The hydraulic end of either centrifugal or axial type is built either as a true pump or a turbine in order to perform well. The motor or generator is a very robust asynchronous induction machine running in air in a waterproof housing. Larger units have a high duty planetary gearbox between the motor/generator and pump/turbine. The shaft end is sealed off by 2 mechanical seals. In between the seals is an oil bath with the purpose of cooling and lubricating the seals. On top of the unit there is a waterproof electrical cable connection box for the power cables and control cable. The units are not bolted against the station structure. The axial flow units are seated against a pipe bottom flange. The centrifugal type is guided down and hooked on a discharge connection. Therefore they can be easily raised for inspection or maintenance. The monitoring equipment indicates a too high winding temperature, bearing temperature and water in motor/-generator housing. Speed is measured on variable speed pumps and hydroturbine generators.

STATION DESIGN

Wether there is a pump or a hydroturbine, the inflow, sidewalls, backwalls and minimum water level over the pump/turbine inlets are of major importance. The pump volutes, propeller pump columns and cylinder gates used with submersible hydroturbine generators are all suppressing vortices and air intrusion. The narrow sidewalls and close backwalls with corner fillets suppresses vortices and distributes an even inflow all around the inlet. Special attention must always be taken if the inflow is not in line with the intake.

In sewage and seawater all non-stainless steel surfaces and submersible units are coated with epoxy tar resin with an antifouling top coating. The submersible units are also protected with zinc anodes.

PROBLEMS?

Foam has been a problem in sewage installations. A sprinkler system to prevent foam formation is going to be tested. An air and foam trap had to be installed in the Hammarby plant just after the distribution flume skiboard.

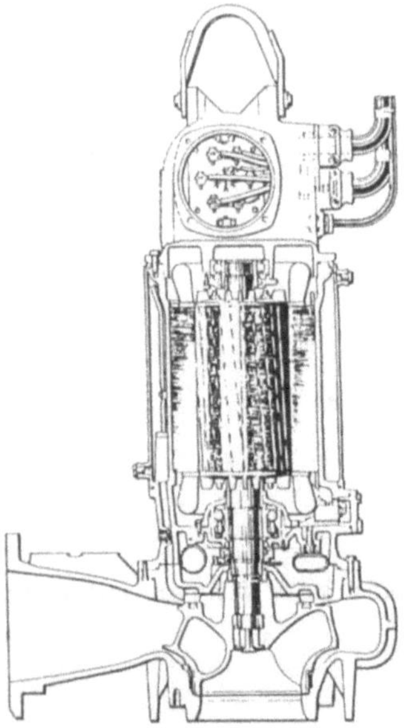

Figure 5. Submersible centrifugal pump direct drive.

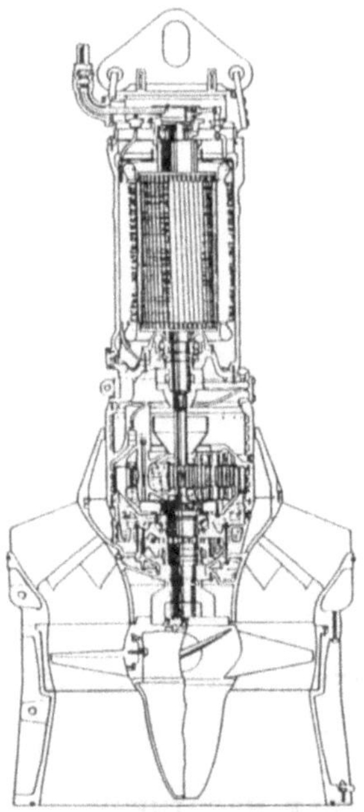

Figure 6. Submersible hydroturbine generator planetary gearbox.

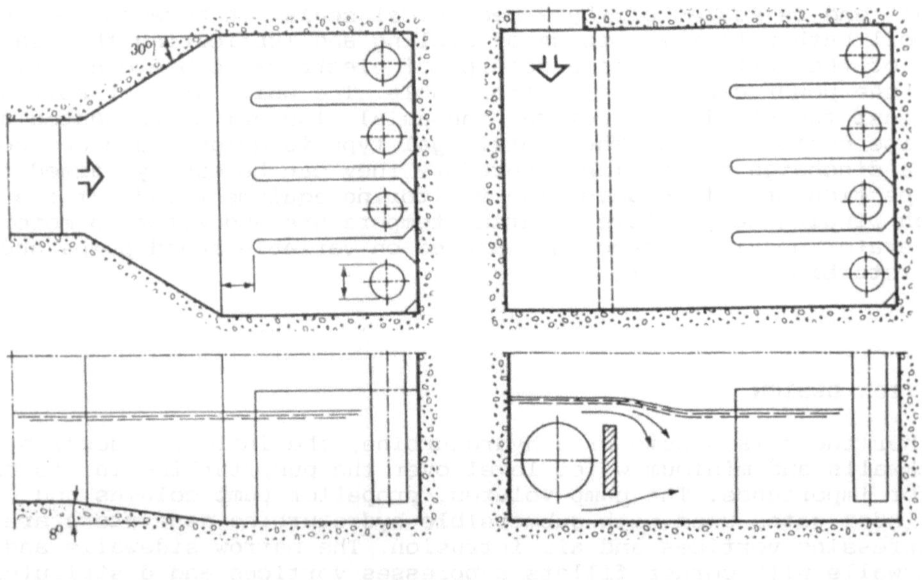

Figure 7. Sump layout for submersible propeller pump - straight and side inflow.

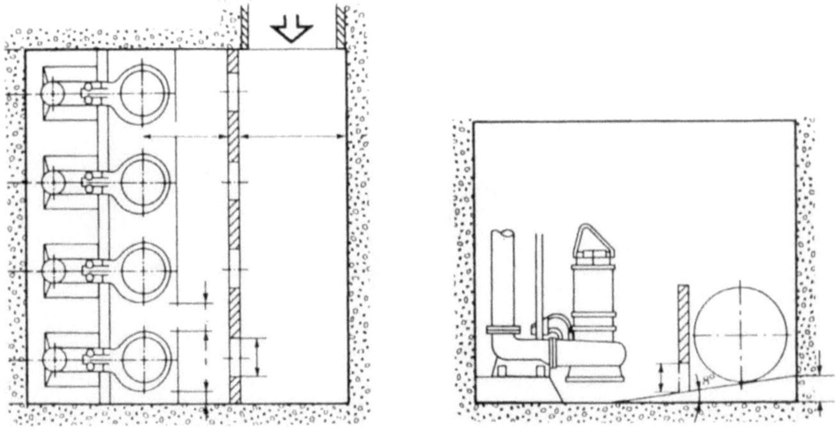

Figure 8. Sump layout for submersible centrifugal pump - side or angle inflow.

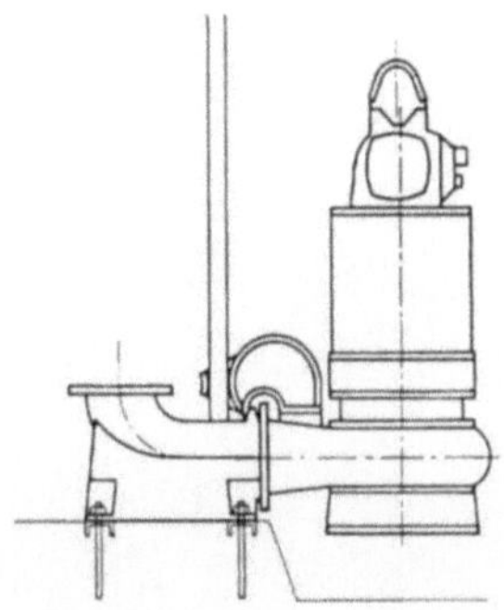

Figure 9. Installation principle submersible centrifugal pump.

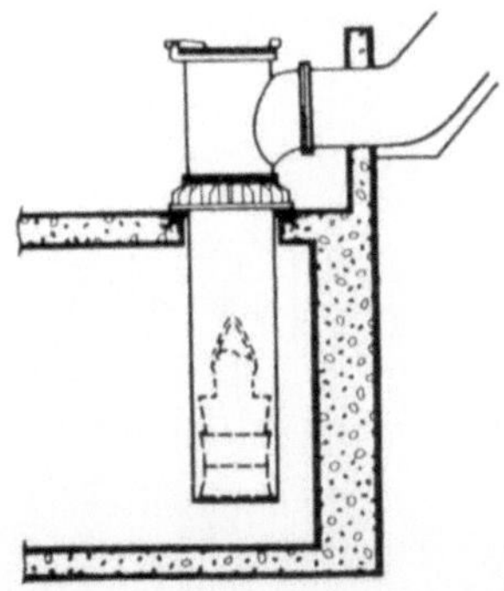

Figure 10. Installation principle submersible propeller pump.

References

* ABB STAL, 1985, 120 MW Heat Pump Station at Solna/Sundbyberg

* Bixio Vincenzo, 1985, New Concepts in the Design of Propeller Pumping Stations, Cleup Padova

* Flygt AB, 1990, Pumping Stations With Large Submersible Centrifugal Pumps

* Flygt AB, 1990, Pumping Stations With Submersible Propeller and Large Low Lift Pumps

* Stjernström - Rune, 1988, Hammarbyverket, värmepumpar, spillvattenpumpar och vattenturbin, KTH DT 1b, VT-88

* Stockholm Energi, 1987, The Hammarby plant

MIX
Papier aus verantwortungsvollen Quellen
Paper from responsible sources
FSC® C105338

If you have any concerns about our products,
you can contact us on
ProductSafety@springernature.com

In case Publisher is established outside the EU,
the EU authorized representative is:
Springer Nature Customer Service Center GmbH
Europaplatz 3, 69115 Heidelberg, Germany

Printed by Libri Plureos GmbH
in Hamburg, Germany